Exponential Sums and their Applications

Mathematics and Its Applications (*Soviet Series*)

Volume 80

Exponential Sums and their Applications

by

N. M. Korobov

Department of Mathematics,
Moscow University,
Moscow, U.S.S.R.

KLUWER ACADEMIC PUBLISHERS
DORDRECHT / BOSTON / LONDON

Library of Congress Cataloging-in-Publication Data

Korobov, N. M. (Nikolaĭ Mikhaĭlovich)
[Trigonometricheskie summy i ikh prilozhenii͡a. English]
Exponential sums and their applications / N.M. Korobov ;
[translated by Yu.N Shakhov].
p. cm. -- (Mathematics and its applications. Soviet series ;
v. 80)
Translation of: Trigonometricheskie summy i ikh prilozhenii͡a.
Includes bibliographical references and index.
ISBN 0-7923-1647-9 (printed on acid free paper)
1. Trigonometric sums. 2. Exponential sums. I. Title.
II. Series: Mathematics and its applications (Kluwer Academic
Publishers). Soviet series ; v. 80.
QA246.8.T75K6713 1992
512'.73--dc20 92-1223

ISBN 0-7923-1647-9

Published by Kluwer Academic Publishers,
P.O. Box 17, 3300 AA Dordrecht, The Netherlands.

Kluwer Academic Publishers incorporates
the publishing programmes of
D. Reidel, Martinus Nijhoff, Dr W. Junk and MTP Press.

Sold and distributed in the U.S.A. and Canada
by Kluwer Academic Publishers,
101 Philip Drive, Norwell, MA 02061, U.S.A.

In all other countries, sold and distributed
by Kluwer Academic Publishers Group,
P.O. Box 322, 3300 AH Dordrecht, The Netherlands.

Printed on acid-free paper

Translated by Yu. N. Shakhov

This book is the translation of the original work
Trigonometrical Sums and their Applications

Printed in the Netherlands

SERIES EDITOR'S PREFACE

'Et moi, ..., si j'avait su comment en revenir, je n'y serais point allé.'
Jules Verne

The series is divergent; therefore we may be able to do something with it.
O. Heaviside

One service mathematics has rendered the human race. It has put common sense back where it belongs, on the topmost shelf next to the dusty canister labelled 'discarded nonsense'.
Eric T. Bell

Mathematics is a tool for thought. A highly necessary tool in a world where both feedback and nonlinearities abound. Similarly, all kinds of parts of mathematics serve as tools for other parts and for other sciences.

Applying a simple rewriting rule to the quote on the right above one finds such statements as: 'One service topology has rendered mathematical physics ...'; 'One service logic has rendered computer science ...'; 'One service category theory has rendered mathematics ...'. All arguably true. And all statements obtainable this way form part of the raison d'être of this series.

This series, *Mathematics and Its Applications*, started in 1977. Now that over one hundred volumes have appeared it seems opportune to reexamine its scope. At the time I wrote

> "Growing specialization and diversification have brought a host of monographs and textbooks on increasingly specialized topics. However, the 'tree' of knowledge of mathematics and related fields does not grow only by putting forth new branches. It also happens, quite often in fact, that branches which were thought to be completely disparate are suddenly seen to be related. Further, the kind and level of sophistication of mathematics applied in various sciences has changed drastically in recent years: measure theory is used (non-trivially) in regional and theoretical economics; algebraic geometry interacts with physics; the Minkowsky lemma, coding theory and the structure of water meet one another in packing and covering theory; quantum fields, crystal defects and mathematical programming profit from homotopy theory; Lie algebras are relevant to filtering; and prediction and electrical engineering can use Stein spaces. And in addition to this there are such new emerging subdisciplines as 'experimental mathematics', 'CFD', 'completely integrable systems', 'chaos, synergetics and large-scale order', which are almost impossible to fit into the existing classification schemes. They draw upon widely different sections of mathematics."

By and large, all this still applies today. It is still true that at first sight mathematics seems rather fragmented and that to find, see, and exploit the deeper underlying interrelations more effort is needed and so are books that can help mathematicians and scientists do so. Accordingly MIA will continue to try to make such books available.

If anything, the description I gave in 1977 is now an understatement. To the examples of interaction areas one should add string theory where Riemann surfaces, algebraic geometry, modular functions, knots, quantum field theory, Kac-Moody algebras, monstrous moonshine (and more) all come together. And to the examples of things which can be usefully applied let me add the topic 'finite geometry'; a combination of words which sounds like it might not even exist, let alone be applicable. And yet it is being applied: to statistics via designs, to radar/sonar detection arrays (via finite projective planes), and to bus connections of VLSI chips (via difference sets). There seems to be no part of (so-called pure) mathematics that is not in immediate danger of being applied. And, accordingly, the applied mathematician needs to be aware of much more. Besides analysis and numerics, the traditional workhorses, he may need all kinds of combinatorics, algebra, probability, and so on.

In addition, the applied scientist needs to cope increasingly with the nonlinear world and the extra mathematical sophistication that this requires. For that is where the rewards are. Linear models are honest and a bit sad and depressing: proportional efforts and results. It is in the nonlinear world that infinitesimal inputs may result in macroscopic outputs (or vice versa). To appreciate what I am hinting at: if electronics were linear we would have no fun with transistors and computers; we would have no TV; in fact you would not be reading these lines.

There is also no safety in ignoring such outlandish things as nonstandard analysis, superspace and anticommuting integration, p-adic and ultrametric space. All three have applications in both electrical engineering and physics. Once, complex numbers were equally outlandish, but they frequently proved the shortest path between 'real' results. Similarly, the first two topics named have already provided a number of 'wormhole' paths. There is no telling where all this is leading - fortunately.

Thus the original scope of the series, which for various (sound) reasons now comprises five subseries: white (Japan), yellow (China), red (USSR), blue (Eastern Europe), and green (everything else), still applies. It has been enlarged a bit to include books treating of the tools from one subdiscipline which are used in others. Thus the series still aims at books dealing with:

- a central concept which plays an important role in several different mathematical and/or scientific specialization areas;
- new applications of the results and ideas from one area of scientific endeavour into another;
- influences which the results, problems and concepts of one field of enquiry have, and have had, on the development of another.

The method of exponential sums is one of the few general methods in (analytic and 'elementary') number theory. It is also, without a doubt, one of the more powerful ones. Getting acquainted with it, and learning to appreciate its ideas and applicability, is a bit of a problem though. The standard sources were composed by and for expert analytic number theorists.

The present monograph gives a straightforward accessible account of the theory with a number of illustrative applications (to number theory, but also to numerical questions). At the same time it contains some new results (in theory) and new applications due to the author.

The main aim of this series is to improve understanding between different mathematical specialisms. In my opinion this book contributes nontrivially to that.

The shortest path between two truths in the real domain passes through the complex domain.

J. Hadamard

La physique ne nous donne pas seulement l'occasion de résoudre des problèmes ... elle nous fait pressentir la solution.

H. Poincaré

Never lend books, for no one ever returns them; the only books I have in my library are books that other folk have lent me.

Anatole France

The function of an expert is not to be more right than other people, but to be wrong for more sophisticated reasons.

David Butler

Bussum, 9 February 1992

Michiel Hazewinkel

CONTENTS

PREFACE

The method of exponential sums is one of a few general methods enabling us to solve a wide range of miscellaneous problems from the theory of numbers and its applications. The strongest results have been obtained with the aid of this method. Therefore knowledge of the fundamentals of the theory of exponential sums is necessary for studying modern number theory.

The study of the method of exponential sums is complicated by the fact that the well-known monographs [44], [16] and [17] are intended for experts, embrace a large number of the fundamental problems at once, are written briefly and for these reasons are not really suitable for a first acquaintance with the subject.

The main aim of the present monograph is to present an as simple as possible exposition of the fundamentals of the theory and, with a series of examples, to show how exponential sums arise and are applied in problems of number theory and in questions connected with their applications. First of all, the book is intended for those who are beginning a study of exponential sums. At the same time, it can be interesting for specialists also, because it contains some results which are not included in other monographs.

This book represents an expanded course of the lectures delivered by the author at the Mechanics and Mathematics Department of Moscow University during the course of many years. It contains the classical results of Gauss, and the methods of Weyl, Mordell and Vinogradov, which are exposed in detail; the traditional applications of exponential sums to the distribution of fractional parts, the estimation of the Riemann zeta-function, the theory of congruences and Diophantine equations are considered too. Some new applications of exponential sums are also included in the book. In particular, questions relating to the distribution of digits in periodic fractions, arising in the expansion of rational numbers under an arbitrary base notation, are considered, and a number of results concerning the completely uniform distribution of fractional parts and the approximate computation of multiple integrals are discussed.

Questions concerning the additive theory of numbers are not included in the book, because for their real understanding one should master the fundamentals of the theory of exponential sums. It will be easier to become acquainted with these and other questions exposed in the monographs [44], [17], [47], [6] and [43] following a subsequent, more profound study of the subject.

To read this book it is sufficient to know the fundamentals of mathematical analysis and to have a knowledge of elementary number theory. For those, who are coming to grips with the subject for the first time, it is recommended to combine the reading of this book with solving problems concerning the investigation and application of the simplest exponential sums [45].

INTRODUCTION

An *exponential sum* is defined as a sum of the form

$$S(P) = \sum_{x} e^{2\pi i\, f(x)}, \tag{1}$$

where x runs over all integers (or some of them) from a certain interval, P is the number of the summands and $f(x)$ is an arbitrary function taking on real values under integer x. Many problems of the number theory and its applications can be reduced to the study of such sums.

Let us show, for instance, how exponential sums arise in solving the problem of possibility to represent a natural number N in the form of a sum of integer powers of natural numbers, the exponents being equal,

$$N = x_1^n + \ldots + x_k^n \tag{2}$$

(Waring's problem). Let n and k be fixed positive integers, P the greatest integer not exceeding $N^{\frac{1}{n}}$ and $T_k(N)$ the number of solutions of the equation (2). For an integer a, let the function $\psi(a)$ be defined by means of the equality

$$\psi(a) = \int_0^1 e^{2\pi i\, a\alpha} d\alpha = \begin{cases} 1 & \text{if} \quad a = 0, \\ 0 & \text{if} \quad a \neq 0. \end{cases}$$

Then obviously

$$T_k(N) = \sum_{x_1,\ldots,x_k=1}^{P} \psi(x_1^n + \ldots + x_k^n - N) = \sum_{x_1,\ldots,x_k=1}^{P} \int_0^1 e^{2\pi i\,(x_1^n+\ldots+x_k^n-N)\alpha} d\alpha$$

$$= \int_0^1 e^{-2\pi i\,\alpha N} \left(\sum_{x=1}^{P} e^{2\pi i\, \alpha x^n} \right)^k d\alpha.$$

Thus the arithmetic problem concerning the number of solutions of the equation (2) is reduced to the study of integral depending on the power of the exponential sum

$$S(P) = \sum_{x=1}^{P} e^{2\pi i\, \alpha x^n}. \tag{3}$$

For applications, the most important sums are those, for which the function $f(x)$ is a polynomial and the summation domain is an interval:

$$S(P) = \sum_{x=Q+1}^{Q+P} e^{2\pi i\, f(x)}, \qquad f(x) = \alpha_1 x + \ldots + \alpha_n x^n. \tag{4}$$

Such exponential sums are called *Weyl's sums* and the degree of the polynomial $f(x)$ *the degree of the Weyl's sum*. So, for example, the sum (3), arising in Waring's problem, is a Weyl's sum of degree n.

The main problem of the theory of exponential sums is to obtain an upper estimate of the modulus of an exponential sum as sharp as possible. As the modulus of every addend of the sum is equal to unity, so for any sum (1), the following trivial estimate is valid:

$$|S(P)| \leqslant P.$$

The first general nontrivial estimates were given by H. Weyl [49]. Under certain requirements for the leading coefficient of the polynomial $f(x)$, he showed that under any ε from the interval $0 < \varepsilon < 1$ there holds the estimate

$$\left| \sum_{x=1}^{P} e^{2\pi i\, f(x)} \right| \leqslant C(n,\varepsilon) P^{1-\frac{\gamma}{2^{n-1}}}, \tag{5}$$

where $\gamma = 1 - \varepsilon$ and $C(n,\varepsilon)$ does not depend on P. Under $n \geqslant 12$ the essential improvement of this result was obtained by I. M. Vinogradov [44], who showed that in the estimate (5) under certain $\gamma > 0$ the right-hand side $C(n,\varepsilon)P^{1-\frac{\gamma}{2^{n-1}}}$ might be replaced by the quantity $C(n)P^{1-\frac{\gamma}{n^2 \log n}}$.

If fractional parts of function $f(x)$ have an integer period, i.e., if under a certain positive integer τ the equality $\{f(x+\tau)\} = \{f(x)\}$, where $\{f(x)\}$ is the fractional part of the function $f(x)$, holds for any integer x, then the sum

$$S(\tau) = \sum_{x=1}^{\tau} e^{2\pi i\, f(x)}$$

is called a *complete exponential sum*. As an example of a complete exponential sum we can take the Weyl's sum, in which all coefficients of the polynomial $f(x)$ are rational and the number of summands is equal to the common denominator of the coefficients:

$$S(q) = \sum_{x=1}^{q} e^{2\pi i \frac{a_1 x + \ldots + a_n x^n}{q}}. \tag{6}$$

Under $a_n \not\equiv 0 \pmod q$ such sums are called *complete rational sums of degree n*. There are more precise estimates of these sums, than estimates of Weyl's sums of the general form.

The thorough research of complete rational sums of the second degree was carried out by Gauss. In particular, he showed that under $(a,q)=1$ for the modulus of the sum

$$S(q)=\sum_{x=1}^{q} e^{2\pi i\,\frac{ax^2}{q}}$$

the equalities

$$|S(q)|=\begin{cases}\sqrt{q} & \text{if}\quad q\equiv 1 \pmod 2,\\ \sqrt{2q} & \text{if}\quad q\equiv 0 \pmod 4,\\ 0 & \text{if}\quad q\equiv 2 \pmod 4)\end{cases}$$

are valid.

For complete rational sums of an arbitrary degree under a prime q Mordell [36] obtained the estimate

$$\left|\sum_{x=1}^{q} e^{2\pi i\,\frac{a_1x+\ldots+a_nx^n}{q}}\right| \leqslant C(n)q^{1-\frac{1}{n}}, \tag{7}$$

where $C(n)$ does not depend on q. Hua Loo-Keng [17] extended this estimation to the case of an arbitrary positive integer q. An essential improvement of the Mordell's result was got by A. Weil [48], who showed that under a prime q the modulus of the sum (7) does not exceed the quantity $(n-1)\sqrt{q}$. Under fixed n and increasing q the estimates by A. Weil and Hua Loo-Keng are the best possible, apart from the values of the constants, and do not admit further essential improvement.

Another example of complete sums, different from the complete rational sum (6), is a sum with exponential function

$$S(\tau)=\sum_{x=1}^{\tau} e^{2\pi i\,\frac{aq^x}{m}}, \tag{8}$$

where $(q,m)=1$ and τ is the order of q for modulus m. The problem of the number of occurrences of a fixed block of digits in the first P digits of a periodical fraction, arising under q-adic expansion of an arbitrary rational number $\frac{b}{m}$, is reduced to estimations of sums (8) and sums $S(P)$ for $P\leqslant\tau$ [32]. The magnitude of the sum (8) depends on the characterization of prime factorization of m and it turns out that for complete sums this magnitude is equal to zero in most cases. But if $P<\tau$, then under $n=\frac{\log m}{\log P}$ and m being equal to a power of a prime, the estimate

$$\left|\sum_{x=1}^{P} e^{2\pi i\,\frac{aq^x}{m}}\right| \leqslant CP^{1-\frac{\gamma}{n^2}},$$

where C and γ are absolute constants, holds.

The necessity to estimate exponential sums arises in the problem of approximate computation of integrals of an arbitrary multiplicity [23] as well. Let us consider, for instance, a quadrature formula constructed by means of an arbitrary net $M_k = M(\xi_1(k), \xi_2(k))$ $(k = 1, 2, \ldots, P)$

$$\int_0^1\int_0^1 F(x_1, x_2)\, dx_1 dx_2 = \frac{1}{P}\sum_{k=1}^{P} F(\xi_1(k), \xi_2(k)) - R_P[F], \tag{9}$$

where $F(x_1, x_2)$ is a periodic function given by its absolutely convergent Fourier expansion

$$F(x_1, x_2) = \sum_{m_1, m_2=-\infty}^{\infty} C(m_1, m_2) e^{2\pi i (m_1 x_1 + m_2 x_2)}.$$

Substituting the series into equality (9) we get after interchanging the order of summation

$$R_P[F] = \frac{1}{P} \sum_{m_1, m_2=-\infty}^{\infty}{}' C(m_1, m_2) \sum_{k=1}^{P} e^{2\pi i (m_1 \xi_1(k) + m_2 \xi_2(k))},$$

where $\sum'$ denotes the summation over all $(m_1, m_2) \neq (0, 0)$. Hence the error term in the quadrature formula (9) satisfies

$$|R_P[F]| \leqslant \frac{1}{P} \sum_{m_1, m_2=-\infty}^{\infty}{}' |C(m_1, m_2)|\, |S(m_1, m_2)|,$$

where the exponential sum

$$S(m_1, m_2) = \sum_{k=1}^{P} e^{2\pi i (m_1 \xi_1(k) + m_2 \xi_2(k))}$$

is determined by the introduction of the net $M(\xi_1(k), \xi_2(k))$. Choosing the functions $\xi_1(k)$ and $\xi_2(k)$ so that the sums $S(m_1, m_2)$ could be estimated sufficiently well, we get the opportunity to construct quadrature formulas of high precision.

Chapter I of this book contains a detailed exposition of some elementary knowledge from the theory of complete exponential sums and sums, which estimations are reduced to estimations of complete sums. Theorems treated in the chapter are comparatively simple, but they constitute the base of the theory of exponential sums of the general form and serve as a necessary preparation to more complicated constructions of Chapter II. To illustrate possible applications of complete sums, the solution of the problem concerning the distribution of digits in the period of fractions, arising in representing rational numbers under an arbitrary base notation, is given in Chapter I.

A technique used in Chapter II is much more complicated than in Chapter I. Chapter II is devoted to an exposition of the theory of Weyl's sums of the general form.

In the chapter, the fundamental methods by Weyl and Vinogradov are presented as well as researches based on the repeated application of the mean value theorem; their applications to estimation of sums, arising in the Riemann zeta-function theory [25]–[28], are given also.

In Chapter III, the exponential sums applications to the distribution of fractional parts and the construction of quadrature formulas are considered. The Weyl theory of uniform distribution is exposed, the questions of complete uniform distribution [20] and their connection with the theory of normal numbers [22] are also considered there. The final part of the chapter is devoted to the problem of approximate calculation of multiple integrals and to construction of interpolation formulas for functions of many variables [23], [29], and [30].

CHAPTER I

COMPLETE EXPONENTIAL SUMS

§ 1. Sums of the first degree

The simplest example of Weyl's sums is the sum of the first degree

$$S(P) = \sum_{x=Q+1}^{Q+P} e^{2\pi i\,\alpha x}.$$

This sum pertains to a number of a few exponential sums, which can be not only estimated but evaluated immediately. In fact, if α is an integer, then $e^{2\pi i\,\alpha} = 1$ and therefore

$$\sum_{x=Q+1}^{Q+P} e^{2\pi i\,\alpha x} = P.$$

But if α is not an integer, then $e^{2\pi i\,\alpha} \neq 1$, and, summing the geometric progression, we have

$$\sum_{x=Q+1}^{Q+P} e^{2\pi i\,\alpha x} = \frac{e^{2\pi i\,\alpha P} - 1}{e^{2\pi i\,\alpha} - 1} e^{2\pi i\,\alpha(Q+1)}. \tag{10}$$

But usually it is more convenient to use not these exact equalities but the following estimate:

LEMMA 1. *Let α be an arbitrary real number, Q an integer, and P a positive integer. Then*

$$\left| \sum_{x=Q+1}^{Q+P} e^{2\pi i\,\alpha x} \right| \leqslant \min\left(P, \frac{1}{2\|\alpha\|}\right), \tag{11}$$

where $\|\alpha\|$ is the distance from α to the nearest integer.

Proof. Since the both sides of (11) are even periodic functions of α with period 1, then it suffices to prove the estimate (11) for $0 \leqslant \alpha \leqslant \frac{1}{2}$. Observing that over this interval

$$\left|e^{2\pi i\,\alpha} - 1\right| = 2\sin \pi\alpha \geqslant 4\alpha = 4\|\alpha\|,$$

then under $\alpha \neq 0$ from the equality (10) we get

$$\left|\sum_{x=Q+1}^{Q+P} e^{2\pi i\,\alpha x}\right| = \frac{|e^{2\pi i\,\alpha P} - 1|}{|e^{2\pi i\,\alpha} - 1|} \leqslant \frac{1}{2\|\alpha\|}.$$

For $\frac{1}{2P} \leqslant \alpha \leqslant \frac{1}{2}$ using this estimate and for $0 \leqslant \alpha < \frac{1}{2P}$ applying the trivial estimate

$$\left|\sum_{x=Q+1}^{Q+P} e^{2\pi i\,\alpha x}\right| \leqslant P,$$

we obtain the assertion of the lemma.

Let a be an arbitrary integer and q a positive integer. We define the function $\delta_q(a)$ with the help of the equality

$$\delta_q(a) = \begin{cases} 1 & \text{if } \ a \equiv 0 \pmod q, \\ 0 & \text{if } \ a \not\equiv 0 \pmod q. \end{cases}$$

In the next lemma the connection between this function and complete rational sums of the first degree will be established.

LEMMA 2. *For any integer a and any positive integer q we have the equality*

$$\delta_q(a) = \frac{1}{q}\sum_{x=1}^{q} e^{2\pi i\,\frac{ax}{q}}. \tag{12}$$

Proof. If $a \equiv 0 \pmod q$, then

$$\frac{1}{q}\sum_{x=1}^{q} e^{2\pi i\,\frac{ax}{q}} = \frac{1}{q}\sum_{x=1}^{q} 1 = 1.$$

Now let $a \not\equiv 0 \pmod q$. Then we get

$$\frac{1}{q}\sum_{x=1}^{q} e^{2\pi i\,\frac{ax}{q}} = \frac{1}{q}\,\frac{e^{2\pi i\,a} - 1}{e^{2\pi i\,\frac{a}{q}} - 1}\,e^{2\pi i\,\frac{a}{q}} = 0.$$

The assertion of the lemma obviously follows from these equalities and the definition of $\delta_q(a)$.

The function $\delta_q(x)$ will be used in the further exposition permanently. Its importance is determined by the fact that it enables us to establish the connection between the exponential sums' investigation and the question of the number of solutions of congruences.

Let us consider, for instance, the question of the number of solutions of the congruence

$$x_1^n + \ldots + x_k^n \equiv \lambda \pmod q, \tag{13}$$

that is analogous to the question of the number of solutions of Waring's equation (2), which was mentioned in the introduction. We denote the number of solutions of this congruence, as the variables $x_1, \ldots, x_k$ run through complete sets of residues to modulus q independently, by $T(\lambda)$. Obviously, by virtue of the definition of the function $\delta_q(x)$

$$T(\lambda) = \sum_{x_1,\ldots,x_k=1}^{q} \delta_q(x_1^n + \ldots + x_k^n - \lambda).$$

Hence it follows by Lemma 2 that

$$T(\lambda) = \sum_{x_1,\ldots,x_k=1}^{q} \frac{1}{q} \sum_{a=1}^{q} e^{2\pi i \frac{a(x_1^n+\ldots+x_k^n-\lambda)}{q}}$$

$$= \frac{1}{q} \sum_{a=1}^{q} e^{-2\pi i \frac{a\lambda}{q}} \sum_{x_1,\ldots,x_k=1}^{q} e^{2\pi i \frac{a(x_1^n+\ldots+x_k^n)}{q}}$$

$$= \frac{1}{q} \sum_{a=1}^{q} e^{-2\pi i \frac{a\lambda}{q}} \left(\sum_{x=1}^{q} e^{2\pi i \frac{ax^n}{q}} \right)^k .$$

Thus the number of solutions of the congruence (13) is represented in terms of complete rational exponential sums

$$S(a, q) = \sum_{x=1}^{q} e^{2\pi i \frac{ax^n}{q}} .$$

We expose some properties of the function $\delta_q(x)$, which follow from its definition immediately.

1°. The function $\delta_q(x)$ is periodic. Its period is equal to q.

2°. If $(a, q) = 1$ and b is an arbitrary integer, then the equalities

$$\delta_q(ax) = \delta_q(x),$$

$$\sum_{x=1}^{q} \delta_q(ax + b) = 1$$

are valid.

3°. Under any positive integer q_1, the equalities

$$\delta_{q_1 q}(q_1 x) = \delta_q(x), \qquad \sum_{y=1}^{q_1} \delta_{q_1 q}(x + qy) = \delta_q(x)$$

hold.

4°. If $(q_1, q) = 1$, then the equality

$$\delta_{q_1 q}(x) = \delta_{q_1}(x)\delta_q(x)$$

is valid.

5°. Under any positive integer P, which does not exceed q, we have

$$\sum_{y=1}^{P} \delta_q(x-y) = \begin{cases} 1 & \text{if} \quad 1 \leqslant x \leqslant P, \\ 0 & \text{if} \quad P < x \leqslant q. \end{cases} \tag{14}$$

LEMMA 3. *Let q be an arbitrary positive integer, $1 \leqslant a < q$, and $(a, q) = 1$. Then the estimates*

$$\sum_{x=1}^{q-1} \frac{1}{\left\| \frac{ax}{q} \right\|} \leqslant 2\,q \log q,$$

$$\sum_{x=1}^{q-1} \frac{1}{x\left\| \frac{ax}{q} \right\|} \leqslant 18\,M \log^2 q,$$

where M is the largest among the partial quotients of the simple continued fraction of the number $\frac{a}{q}$, hold.

Proof. Let m be an arbitrary positive integer. Under $x \geqslant 1$ using the inequality

$$\frac{1}{x} \leqslant \log(2x+1) - \log(2x-1),$$

we obtain

$$\sum_{1 \leqslant x \leqslant m} \frac{1}{x} \leqslant \sum_{1 \leqslant x \leqslant m} \log(2x+1) - \sum_{1 \leqslant x \leqslant m} \log(2x-1) = \log(2m+1).$$

Hence under odd and even q, respectively, it follows that

$$\sum_{x=1}^{\frac{q-1}{2}} \frac{1}{x} \leqslant \log q, \qquad \sum_{x=1}^{\frac{q-2}{2}} \frac{1}{x} \leqslant \log(q-1) \leqslant -\frac{1}{q} + \log q. \tag{15}$$

Since the function $\|\frac{ax}{q}\|$ is periodic with period q and $(a, q) = 1$, then under odd q according to (15) we get

$$\sum_{x=1}^{q-1} \frac{1}{\left\| \frac{ax}{q} \right\|} = \sum_{x=1}^{q-1} \frac{1}{\left\| \frac{x}{q} \right\|} = 2 \sum_{x=1}^{\frac{q-1}{2}} \frac{1}{\left\| \frac{x}{q} \right\|} = 2\,q \sum_{x=1}^{\frac{q-1}{2}} \frac{1}{x} \leqslant 2\,q \log q.$$

But the same estimate is obtained by (15) under even q as well:

$$\sum_{x=1}^{q-1} \frac{1}{\left\| \frac{ax}{q} \right\|} = 2 + 2q \sum_{x=1}^{\frac{q-2}{2}} \frac{1}{x} \leqslant 2q \log q.$$

The first assertion of the lemma is proved.

To prove the second assertion we shall apply the Abel summation formula

$$\sum_{x=1}^{q-1} u_x v_x = u_q \sum_{x=1}^{q-1} v_x + \sum_{m=1}^{q-1} \left(u_m - u_{m+1}\right) \sum_{x=1}^{m} v_x.$$

Under $u_x = \frac{1}{x}$ and $v_x = \frac{1}{\|\frac{ax}{q}\|}$ we obtain

$$\sum_{x=1}^{q-1} \frac{1}{x\left\| \frac{ax}{q} \right\|} = \frac{1}{q} \sum_{x=1}^{q-1} \frac{1}{\left\| \frac{ax}{q} \right\|} + \sum_{m=1}^{q-1} \frac{1}{m(m+1)} \sum_{x=1}^{m} \frac{1}{\left\| \frac{ax}{q} \right\|}. \tag{16}$$

Let the expansion of the number $\frac{a}{q}$ in simple continued fraction be

$$\frac{a}{q} = \cfrac{1}{q_1 + \cfrac{1}{q_2 + \ddots + \cfrac{1}{q_n}}}.$$

Then under $\nu = 1, 2, \ldots, n$ the following equalities take place:

$$\frac{a}{q} = \frac{P_\nu}{Q_\nu} + \frac{\theta_\nu}{Q_\nu^2} \qquad \left(|\theta_\nu| \leqslant 1\right), \tag{17}$$

where P_ν and Q_ν are relatively prime, $1 = Q_0 \leqslant Q_1 < \ldots < Q_n = q$, $Q_\nu \leqslant (q_\nu + 1)Q_{\nu-1} \leqslant 2MQ_{\nu-1}$.

If $1 \leqslant m < \frac{1}{2}q$, then determining ν from the condition

$$\frac{1}{2} Q_{\nu-1} \leqslant m < \frac{1}{2} Q_\nu$$

and using the equality (17), we get

$$\left\| \frac{ax}{q} \right\| = \left\| \frac{P_\nu x}{Q_\nu} + \frac{\theta_\nu x}{Q_\nu^2} \right\| \geqslant \left\| \frac{P_\nu x}{Q_\nu} \right\| - \left\| \frac{\theta_\nu x}{Q_\nu^2} \right\|. \tag{18}$$

Since under $1 \leqslant x < \frac{1}{2}Q_\nu$ we have

$$\left\| \frac{\theta_\nu x}{Q_\nu^2} \right\| \leqslant \frac{1}{2Q_\nu} \leqslant \frac{1}{2} \left\| \frac{P_\nu x}{Q_\nu} \right\|,$$

hence from (18) it follows that

$$\left\| \frac{ax}{q} \right\| \geqslant \frac{1}{2} \left\| \frac{P_\nu x}{Q_\nu} \right\|.$$

Then using the first inequality of the lemma, we obtain

$$\sum_{x=1}^{m} \frac{1}{\left\| \frac{ax}{q} \right\|} \leqslant 2 \sum_{x=1}^{Q_\nu - 1} \frac{1}{\left\| \frac{P_\nu x}{Q_\nu} \right\|} \leqslant 4\, Q_\nu \log Q_\nu$$
$$\leqslant 8\, M Q_{\nu-1} \log q \leqslant 16\, Mm \log q. \tag{19}$$

But if $\frac{1}{2} q \leqslant m < q$, then

$$\sum_{x=1}^{m} \frac{1}{\left\| \frac{ax}{q} \right\|} \leqslant \sum_{x=1}^{q-1} \frac{1}{\left\| \frac{ax}{q} \right\|} \leqslant 2\, q \log q \leqslant 4\, m \log q,$$

and, therefore, the estimate (19) holds not only for $m < \frac{1}{2} q$, but for any $m < q$ as well. Substituting it into the equality (16), we get the second assertion of the lemma:

$$\sum_{x=1}^{q-1} \frac{1}{x \left\| \frac{ax}{q} \right\|} \leqslant 2 \log q + \sum_{m=1}^{q-1} \frac{16\, M \log q}{m+1} \leqslant 18\, M \log^2 q.$$

Now we'll show how these lemmas, containing quite a little information concerning exponential sums, enable us to get nontrivial arithmetic results.

Let $(a, q) = 1$, $P_1 \leqslant q$, $P_2 \leqslant q$, and T be the number of solutions of the congruence

$$a x_1 \equiv x_2 \pmod q, \qquad 1 \leqslant x_1 \leqslant P_1,\ 1 \leqslant x_2 \leqslant P_2. \tag{20}$$

If P_1 or P_2 equals q, then, evidently,

$$T = \frac{1}{q} P_1 P_2.$$

The question becomes more complicated, if both P_1 and P_2 are less than q. In this case, it can be shown that

$$T = \frac{1}{q} P_1 P_2 + 9\, \theta M \log^2 q, \qquad |\theta| \leqslant 1, \tag{21}$$

where M is the largest among partial quotients of the simple continued fraction of the number $\frac{a}{q}$.

Really, using Lemma 2, we obtain

$$T=\sum_{x_1=1}^{P_1}\sum_{x_2=1}^{P_2}\delta_q(ax_1-x_2)=\frac{1}{q}\sum_{x_1=1}^{P_1}\sum_{x_2=1}^{P_2}\sum_{x=1}^{q}e^{2\pi i\,\frac{(ax_1-x_2)x}{q}}.$$

Hence, after singling out the summand with $x=q$, it follows that

$$T=\frac{1}{q}P_1P_2+R, \tag{22}$$

where

$$|R|=\frac{1}{q}\left|\sum_{x=1}^{q-1}\left(\sum_{x_1=1}^{P_1}e^{2\pi i\,\frac{axx_1}{q}}\right)\left(\sum_{x_2=1}^{P_2}e^{-2\pi i\,\frac{xx_2}{q}}\right)\right|$$
$$\leqslant\frac{1}{q}\sum_{x=1}^{q-1}\left|\sum_{x_1=1}^{P_1}e^{2\pi i\,\frac{axx_1}{q}}\right|\left|\sum_{x_2=1}^{P_2}e^{2\pi i\,\frac{xx_2}{q}}\right|.$$

Thus the problem concerning the number of solutions of the congruence (20) is reduced to the problem of the estimation of Weyl's sums of the first degree. Using Lemma 1 and observing that $\|\frac{x}{q}\|$ and $\|\frac{ax}{q}\|$ are even periodic functions with period q, we get

$$|R|\leqslant\frac{1}{q}\sum_{x=1}^{q-1}\min\left(P_1,\frac{1}{2\|\frac{ax}{q}\|}\right)\min\left(P_2,\frac{1}{2\|\frac{x}{q}\|}\right)$$
$$\leqslant\frac{1}{4q}\sum_{1\leqslant|x|\leqslant\frac{1}{2}q}\frac{1}{\|\frac{x}{q}\|\,\|\frac{ax}{q}\|}=\frac{1}{2}\sum_{1\leqslant x\leqslant\frac{1}{2}q}\frac{1}{x\|\frac{ax}{q}\|}.$$

Hence according to Lemma 3 it follows that

$$|R|\leqslant 9\,M\log^2 q,$$

and by (22) this estimate is equivalent to the equality (21).

§ 2. General properties of complete sums

As it was said above, the sum

$$S(\tau)=\sum_{x=1}^{\tau}e^{2\pi i\,f(x)} \tag{23}$$

is called a complete exponential sum, if under any integer x for fractional parts of the function $f(x)$, the equality $\{f(x+\tau)\}=\{f(x)\}$ is satisfied.

We shall expose some examples of complete sums. Let $a_1, \ldots, a_n$ be integers and $\varphi(x) = a_1 x + \ldots + a_n x^n$. Since, obviously,

$$(x+q)^\nu \equiv x^\nu \pmod q \qquad (\nu = 1, 2, \ldots, n),$$

then the following congruences hold:

$$\sum_{\nu=1}^{n} a_\nu (x+q)^\nu \equiv \sum_{\nu=1}^{n} a_\nu x^\nu \pmod q,$$

$$\varphi(x+q) \equiv \varphi(x) \pmod q.$$

But then under any integer x

$$\left\{\frac{\varphi(x+q)}{q}\right\} = \left\{\frac{\varphi(x)}{q}\right\},$$

and, therefore, the sum

$$S(q) = \sum_{x=1}^{q} e^{2\pi i \frac{\varphi(x)}{q}} = \sum_{x=1}^{q} e^{2\pi i \frac{a_1 x + \ldots + a_n x^n}{q}},$$

which was called a complete rational sum in the introduction, is a complete exponential sum in the sense of the definition (23).

Now let us consider a sum with exponential function

$$S(\tau) = \sum_{x=1}^{\tau} e^{2\pi i \frac{aq^x}{m}}, \tag{24}$$

where $(a, m) = 1$, $(q, m) = 1$ and τ is the order of q for modulus m. Let q^{-1} denote the solution of the congruence $qx \equiv 1 \pmod m$. Then using the congruence $q^\tau \equiv 1 \pmod m$, under any integer x we obtain

$$\left\{\frac{aq^{x+\tau}}{m}\right\} = \left\{\frac{aq^x}{m}\right\}.$$

Therefore τ is a period of fractional parts of the function $\frac{aq^x}{m}$ and the sum (24) is a complete exponential sum.

Expose some properties of complete sums, which follow from the definition directly.

1°. The magnitude of the complete exponential sum (23) will not change, if the summation variable runs through any complete set of residues to modulus τ instead of the interval $[1, \tau]$.

Really, since $\{f(x+\tau)\} = \{f(x)\}$, then under $x \equiv y \pmod \tau$ the equality $\{f(x)\} = \{f(y)\}$ holds. But then

$$e^{2\pi i\, f(x)} = e^{2\pi i\, f(y)}$$

and the totality of the summands of the sum (23) is independent of whichever complete set of incongruent residues to modulus τ is run by the summation variable.

2°. If $(\lambda, \tau) = 1$, μ is an integer and n is a positive integer, then for complete sums the equalities

$$\sum_{x=1}^{\tau} e^{2\pi i\, f(x)} = \sum_{x=1}^{\tau} e^{2\pi i\, f(\lambda x + \mu)}, \tag{25}$$

$$\sum_{x=1}^{n\tau} e^{2\pi i\, f(x)} = n \sum_{x=1}^{\tau} e^{2\pi i\, f(x)} \tag{26}$$

hold.

The first among these equalities is a particular case of the property 1°, because under $(\lambda, \tau) = 1$ the linear function $\lambda x + \mu$ runs through a complete set of residues to modulus τ, when x runs through a complete residue set to modulus τ. The second equality follows from 1° as well, for under varying from 1 to $n\tau$ the summation variable runs n times through complete residue set to modulus τ.

3°. If sums

$$\sum_{x=1}^{\tau} e^{2\pi i\, f_1(x)} \quad \text{and} \quad \sum_{x=1}^{\tau} e^{2\pi i\, f_2(x)} \tag{27}$$

are complete, then the sum

$$\sum_{x=1}^{\tau} e^{2\pi i\, (f_1(x) + f_2(x))} \tag{28}$$

is complete also.

Really, it follows from completeness of the sums (27), that fractional parts of the functions $f_1(x)$ and $f_2(x)$ have the same period τ:

$$\{f_1(x+\tau)\} = \{f_1(x)\}, \qquad \{f_2(x+\tau)\} = \{f_2(x)\}.$$

But then

$$\{f_1(x+\tau) + f_2(x+\tau)\} = \{f_1(x) + f_2(x)\}$$

and, therefore, the sum (28) is a complete exponential sum.

THEOREM 1 (multiplication formula). *Let under integers x*

$$\{f(x)\} = \{f_1(x) + \ldots + f_s(x)\}, \tag{29}$$

where fractional parts of the functions $f_1(x), \ldots, f_s(x)$ are periodic and their periods $\tau_1, \ldots, \tau_s$ are relatively prime to each other. Then the equality

$$\sum_{x=1}^{\tau_1 \ldots \tau_s} e^{2\pi i\, f(x)} = \prod_{\nu=1}^{s} \sum_{x_\nu=1}^{\tau_\nu} e^{2\pi i\, f_\nu(x_\nu)} \tag{30}$$

holds.

Proof. Since by the assumption

$$\{f_\nu(x+\tau_\nu)\} = \{f_\nu(x)\} \qquad (\nu = 1,2,\dots,s) \tag{31}$$

and by (29)

$$\{f(x+\tau_1\dots\tau_s)\} = \{f(x)\},$$

then all the exponential sums in the equality (30) are complete. Let variables $x_1,\dots,x_s$ run independently through complete residue sets to moduli $\tau_1,\dots,\tau_s$, respectively. Since the $\tau_1,\dots,\tau_s$ are coprime, then the sum

$$x_1\tau_2\dots\tau_s+\dots+\tau_1\dots\tau_{s-1}x_s$$

runs through a complete residue set to modulus $\tau_1\dots\tau_s$, and, therefore,

$$\sum_{x=1}^{\tau_1\dots\tau_s} e^{2\pi i\, f(x)} = \sum_{x_1=1}^{\tau_1}\dots\sum_{x_s=1}^{\tau_s} e^{2\pi i\, f(x_1\tau_2\dots\tau_s+\dots+\tau_1\dots\tau_{s-1}x_s)}. \tag{32}$$

Since by (29) and (31)

$$\{f(x_1\tau_2\dots\tau_s+\dots+\tau_1\dots\tau_{s-1}x_s)\} = \{f_1(x_1\tau_2\dots\tau_s)+\dots+f_s(\tau_1\dots\tau_{s-1}x_s)\},$$

then the equality (32) may be rewritten in the form

$$\sum_{x=1}^{\tau_1\dots\tau_s} e^{2\pi i\, f(x)} = \sum_{x_1=1}^{\tau_1}\dots\sum_{x_s=1}^{\tau_s} e^{2\pi i\,(f_1(x_1\tau_2\dots\tau_s)+\dots+f_s(\tau_1\dots\tau_{s-1}x_s))}.$$

Hence, using the property (25), we obtain the multiplication formula:

$$\sum_{x=1}^{\tau_1\dots\tau_s} e^{2\pi i\, f(x)} = \sum_{x_1=1}^{\tau_1}\dots\sum_{x_s=1}^{\tau_s} e^{2\pi i\,(f_1(x_1)+\dots+f_s(x_s))} = \prod_{\nu=1}^{s}\sum_{x_\nu=1}^{\tau_\nu} e^{2\pi i\, f_\nu(x_\nu)}.$$

In a number of cases, the multiplication formula simplifies the study of complete sums. As an example of that we shall consider complete rational sums.

Let $\varphi(x) = a_1x+\dots+a_nx^n$ be an arbitrary polynomial with integral coefficients, $q = p_1^{\alpha_1}\dots p_s^{\alpha_s}$ prime factorization of q, and numbers $b_1,\dots,b_n$ be chosen to satisfy the congruence

$$1 \equiv b_1p_2^{\alpha_2}\dots p_s^{\alpha_s}+\dots+p_1^{\alpha_1}\dots p_{s-1}^{\alpha_{s-1}}b_s \pmod q. \tag{33}$$

Then for complete rational sums the following equality holds

$$\sum_{x=1}^{q} e^{2\pi i\frac{\varphi(x)}{q}} = \prod_{\nu=1}^{s}\sum_{x_\nu=1}^{p_\nu^{\alpha_\nu}} e^{2\pi i\frac{b_\nu\varphi(x_\nu)}{p_\nu^{\alpha_\nu}}}. \tag{34}$$

Really, since

$$\left\{\frac{\varphi(x+q)}{q}\right\}=\left\{\frac{\varphi(x)}{q}\right\},\qquad \left\{\frac{b_\nu\varphi(x+p_\nu^{\alpha_\nu})}{p_\nu^{\alpha_\nu}}\right\}=\left\{\frac{b_\nu\varphi(x)}{p_\nu^{\alpha_\nu}}\right\}\qquad (1\leqslant \nu\leqslant s)$$

and by (33)

$$\left\{\frac{\varphi(x)}{q}\right\}=\left\{\frac{b_1\varphi(x)}{p_1^{\alpha_1}}+\ldots+\frac{b_s\varphi(x)}{p_s^{\alpha_s}}\right\},$$

then applying Theorem 1, we obtain the equality (34).

The multiplication formula (34) reduces the investigation of complete rational sums with an arbitrary denominator q to the investigation of simpler sums with a denominator being a power of a prime.

As another example on the multiplication formula we shall prove the equality

$$\sum_{x=1}^{q-1} e^{2\pi i\,\frac{x^2}{q}}=(1-i^q)\sum_{x=1}^{q-1} e^{2\pi i\,\frac{x^2}{4q}},\qquad q\equiv 1 \pmod 2, \tag{35}$$

which will be needed later in studying Gaussian sums. Consider the sum

$$S=\sum_{x=1}^{4q} e^{2\pi i\,\frac{x^2}{4q}}.$$

Single out the summands, for which x is a multiple of q, and group the others in four sums:

$$\begin{aligned} S&=\sum_{x=1}^{4} e^{2\pi i\,\frac{qx^2}{4}}+\sum_{x=1}^{q-1}\left(e^{2\pi i\,\frac{x^2}{4q}}+e^{2\pi i\,\frac{(2q-x)^2}{4q}}+e^{2\pi i\,\frac{(2q+x)^2}{4q}}+e^{2\pi i\,\frac{(4q-x)^2}{4q}}\right)\\ &=\sum_{x=1}^{4} e^{2\pi i\,\frac{qx^2}{4}}+4\sum_{x=1}^{q-1} e^{2\pi i\,\frac{x^2}{4q}}. \end{aligned}\tag{36}$$

On the other hand, according to the multiplication formula

$$S=\sum_{x_1=1}^{4} e^{2\pi i\,\frac{b_1x_1^2}{4}}\sum_{x_2=1}^{q} e^{2\pi i\,\frac{b_2x_2^2}{q}},$$

where b_1 and b_2 satisfy the congruence $qb_1+4b_2\equiv 1 \pmod{4q}$. Since this congruence is satisfied under $b_1=q$ and $b_2=\frac{1}{4}(1-q^2)$, then after singling out the summand with $x_2=q$ and replacing x_2 by $2x$, we obtain

$$\begin{aligned} S&=\sum_{x_1=1}^{4} e^{2\pi i\,\frac{b_1x_1^2}{4}}+\sum_{x_1=1}^{4} e^{2\pi i\,\frac{b_1x_1^2}{4}}\sum_{x=1}^{q-1} e^{2\pi i\,\frac{4b_2x^2}{q}}\\ &=\sum_{x=1}^{4} e^{2\pi i\,\frac{qx^2}{4}}+\sum_{x_1=1}^{4} e^{2\pi i\,\frac{qx_1^2}{4}}\sum_{x=1}^{q-1} e^{2\pi i\,\frac{x^2}{q}}. \end{aligned}\tag{37}$$

Now observing that

$$\sum_{x_1=1}^{4} e^{2\pi i \frac{q x_1^2}{4}} = 2(1+i^q),$$

from (36) and (37) we get the equality (35):

$$\sum_{x=1}^{q-1} e^{2\pi i \frac{x^2}{q}} = \frac{4}{2(1+i^q)} \sum_{x=1}^{q-1} e^{2\pi i \frac{x^2}{4q}} = (1-i^q) \sum_{x=1}^{q-1} e^{2\pi i \frac{x^2}{4q}}.$$

Now we shall consider a certain class of exponential sums, whose nontrivial estimates can be easily obtained by the reduction of the problem to the estimation of complete sums.

Let fractional parts of a function $f(x)$ be periodic, their least period be equal to τ, $1 \leqslant P < \tau$ and Q an arbitrary integer. Then the sum

$$S(P) = \sum_{x=Q+1}^{Q+P} e^{2\pi i f(x)} \tag{38}$$

is called an *incomplete exponential sum.*

THEOREM 2. *For any incomplete exponential sum $S(P)$ defined by the equality (38), the estimate*

$$|S(P)| \leqslant \max_{1 \leqslant a \leqslant \tau} \left| \sum_{x=1}^{\tau} e^{2\pi i (f(x) + \frac{ax}{\tau})} \right| (1 + \log \tau)$$

holds.

Proof. From the property (14) of the function $\delta_q(x)$ it follows that under $P \leqslant \tau$

$$\sum_{y=Q+1}^{Q+P} \delta_\tau(x-y) = \begin{cases} 1 & \text{if } \ Q+1 \leqslant x \leqslant Q+P, \\ 0 & \text{if } \ Q+P < x \leqslant Q+\tau. \end{cases}$$

Applying this discontinuous factor and using Lemma 2, we obtain

$$\begin{aligned} \sum_{x=Q+1}^{Q+P} e^{2\pi i f(x)} &= \sum_{x=Q+1}^{Q+\tau} e^{2\pi i f(x)} \sum_{y=Q+1}^{Q+P} \delta_\tau(x-y) \\ &= \frac{1}{\tau} \sum_{a=1}^{\tau} \left(\sum_{y=Q+1}^{Q+P} e^{-2\pi i \frac{ay}{\tau}} \right) \sum_{x=Q+1}^{Q+\tau} e^{2\pi i \left(f(x) + \frac{ax}{\tau}\right)}. \end{aligned}$$

Since fractional parts of the functions $f(x)$ and $\frac{ax}{\tau}$ have period τ, then by (28) the latter sum in this equality is complete and, therefore,

$$\sum_{x=Q+1}^{Q+P} e^{2\pi i f(x)} = \frac{1}{\tau} \sum_{a=1}^{\tau} \left(\sum_{y=Q+1}^{Q+P} e^{-2\pi i \frac{ay}{\tau}} \right) \sum_{x=1}^{\tau} e^{2\pi i \left(f(x) + \frac{ax}{\tau}\right)}.$$

Hence, using Lemmas 2 and 3, we get the theorem assertion:

$$\begin{aligned}\left|\sum_{x=Q+1}^{Q+P} e^{2\pi i\, f(x)}\right| &\leqslant \frac{1}{\tau}\sum_{a=1}^{\tau}\left|\sum_{x=1}^{\tau} e^{2\pi i\left(f(x)+\frac{ax}{\tau}\right)}\right| \min\left(P, \frac{1}{2\|\frac{a}{\tau}\|}\right)\\ &\leqslant \frac{1}{\tau}\max_{1\leqslant a\leqslant\tau}\left|\sum_{x=1}^{\tau} e^{2\pi i\left(f(x)+\frac{ax}{\tau}\right)}\right|\sum_{a=1}^{\tau}\min\left(P, \frac{1}{2\|\frac{a}{\tau}\|}\right)\\ &\leqslant \max_{1\leqslant a\leqslant\tau}\left|\sum_{x=1}^{\tau} e^{2\pi i\left(f(x)+\frac{ax}{\tau}\right)}\right|(1+\log\tau).\end{aligned}$$

§ 3. Gaussian sums

A *Gaussian sum* is a complete rational exponential sum of the second degree

$$S(q) = \sum_{x=1}^{q} e^{2\pi i\,\frac{ax^2}{q}},$$

where q is an arbitrary positive integer and $(a, q) = 1$. Gaussian sums as well as the first degree sums considered in the first paragraph can be evaluated precisely. We shall start with a comparatively simple question about the evaluation of the modulus of such sums.

THEOREM 3. *For the modulus of the Gaussian sum, the following equalities hold true:*

$$|S(q)| = \begin{cases}\sqrt{q} & \text{if}\ \ q\equiv 1 \pmod 2,\\ \sqrt{2q} & \text{if}\ \ q\equiv 0 \pmod 4,\\ 0 & \text{if}\ \ q\equiv 2 \pmod 4.\end{cases}$$

Proof. Let the complex conjugate of the sum $S(q)$ be denoted by $\overline{S}(q)$. Then we get

$$|S(q)|^2 = \overline{S}(q)S(q) = \sum_{y=1}^{q} e^{-2\pi i\,\frac{ay^2}{q}}\sum_{x=1}^{q} e^{2\pi i\,\frac{ax^2}{q}}.$$

Utilize the second property of complete sums and replace x by $x+y$ in the inner sum. Then after interchanging the order of summation, we obtain

$$|S(q)|^2 = \sum_{x=1}^{q}\sum_{y=1}^{q} e^{2\pi i\,\frac{a(x+y)^2-ay^2}{q}} = \sum_{x=1}^{q} e^{2\pi i\,\frac{ax^2}{q}}\sum_{y=1}^{q} e^{2\pi i\,\frac{2axy}{q}}.$$

Hence by Lemma 2 it follows that

$$|S(q)|^2 = q\sum_{x=1}^{q} e^{2\pi i\frac{ax^2}{q}}\delta_q(2ax). \tag{39}$$

Since a and q are coprime by the statement, then under odd q the only nonzero summand of the right-hand side of this equality is the summand obtained under $x = q$, and therefore

$$|S(q)|^2 = qe^{2\pi i\frac{aq^2}{q}} = q. \tag{40}$$

But if q is even, then in the sum (39) there are two nonzero summands which are obtained under $x = \frac{1}{2}q$ and $x = q$. Therefore, observing that under even q, from $(a,q) = 1$ it follows that a is odd, we get

$$|S(q)|^2 = q\left(e^{2\pi i\frac{aq}{4}} + 1\right) = q\left(e^{2\pi i\frac{q}{4}} + 1\right) = \begin{cases} 2q & \text{if } q \equiv 0 \pmod 4, \\ 0 & \text{if } q \equiv 2 \pmod 4. \end{cases}$$

The theorem assertion follows from this equality and (40).

Note that in the case of odd q, the assertion of Theorem 3 is valid for sums of the general form, too.

Indeed, let us show that under $(2a_2, q) = 1$ the equality

$$\left|\sum_{x=1}^{q} e^{2\pi i\frac{a_1x+a_2x^2}{q}}\right| = \sqrt{q} \tag{41}$$

holds. Choose b satisfying the congruence $2a_2b \equiv a_1 \pmod q$. Then obviously

$$a_1x + a_2x^2 \equiv a_2(x+b)^2 - a_2b^2 \pmod q$$

and, therefore,

$$\sum_{x=1}^{q} e^{2\pi i\frac{a_1x+a_2x^2}{q}} = e^{-2\pi i\frac{a_2b^2}{q}}\sum_{x=1}^{q} e^{2\pi i\frac{a_2(x+b)^2}{q}}.$$

Hence we obtain the equality (41):

$$\left|\sum_{x=1}^{q} e^{2\pi i\frac{a_1x+a_2x^2}{q}}\right| = \left|\sum_{x=1}^{q} e^{2\pi i\frac{a_2(x+b)^2}{q}}\right| = \sqrt{q}.$$

Let as consider the simplest properties of Gaussian sums. We shall assume that $q = p$, where $p > 2$ is a prime. It is easy to show that under $a = 0 \pmod p$ the following equality holds:

$$\sum_{x=1}^{p} e^{2\pi i\frac{ax^2}{p}} = \sum_{x=1}^{p-1}\left(\frac{x}{p}\right)e^{2\pi i\frac{ax}{p}}, \tag{42}$$

where $\left(\frac{x}{p}\right)$ is Legendre's symbol. Indeed, if x varies from 1 to $p-1$, then x^2 runs twice through values of quadratic residues of p, and since

$$1+\left(\frac{x}{p}\right)=\begin{cases}2 & \text{if } x \text{ is a quadratic residue,}\\ 0 & \text{if } x \text{ is a quadratic non-residue,}\end{cases}$$

then

$$\sum_{x=1}^{p} e^{2\pi i \frac{ax^2}{p}} = 1+\sum_{x=1}^{p-1} e^{2\pi i \frac{ax^2}{p}} = 1+\sum_{x=1}^{p-1}\left[1+\left(\frac{x}{p}\right)\right] e^{2\pi i \frac{ax}{p}}.$$

Hence observing that by Lemma 2 under $a = 0 \pmod p$

$$1+\sum_{x=1}^{p-1} e^{2\pi i \frac{ax}{p}} = p\delta_p(a) = 0,$$

we obtain the equality (42).

Now we shall show that under $a = 0 \pmod p$

$$\sum_{x=1}^{p} e^{2\pi i \frac{ax^2}{p}} = \left(\frac{a}{p}\right)\sum_{x=1}^{p} e^{2\pi i \frac{x^2}{p}}. \tag{43}$$

Indeed, multiplying the equality (42) by $\left(\frac{a^2}{p}\right) = 1$ and observing that ax runs through a complete set of residues prime to p when x runs through such a set, we get

$$\sum_{x=1}^{p} e^{2\pi i \frac{ax^2}{p}} = \left(\frac{a}{p}\right)\sum_{x=1}^{p-1}\left(\frac{ax}{p}\right) e^{2\pi i \frac{ax}{p}} = \left(\frac{a}{p}\right)\sum_{x=1}^{p-1}\left(\frac{x}{p}\right) e^{2\pi i \frac{x}{p}}.$$

The equality (43) follows, because by (42)

$$\sum_{x=1}^{p-1}\left(\frac{x}{p}\right) e^{2\pi i \frac{x}{p}} = \sum_{x=1}^{p} e^{2\pi i \frac{x^2}{p}}.$$

Next we shall show that knowing the modulus of a Gaussian sum it is easy to evaluate its value to within the accuracy of the sign. Indeed, let

$$S(p) = \sum_{x=1}^{p} e^{2\pi i \frac{x^2}{p}}.$$

Then, using the equality (43), we get

$$\overline{S}(p) = \sum_{x=1}^{p} e^{-2\pi i \frac{x^2}{p}} = \left(\frac{-1}{p}\right)\sum_{x=1}^{p} e^{2\pi i \frac{x^2}{p}} = \left(\frac{-1}{p}\right) S(p).$$

Hence after multiplying by $\left(\frac{-1}{p}\right)S(p)$ it follows that

$$S^2(p) = \left(\frac{-1}{p}\right)|S(p)|^2 = \left(\frac{-1}{p}\right)p.$$

Now, since $\left(\frac{-1}{p}\right)$ takes on the value 1 under $p \equiv 1 \pmod 4$ and the value -1 under $p \equiv 3 \pmod 4$, we obtain

$$S(p) = \begin{cases} \pm\sqrt{p} & \text{if } \ p \equiv 1 \pmod 4, \\ \pm i\sqrt{p} & \text{if } \ p \equiv 3 \pmod 4. \end{cases} \tag{44}$$

The question about choosing the proper sign in these equalities is more difficult. Its solution was found by Gauss. A comparatively simple proof of the Gauss theorem given in the paper [9] is exposed below.

THEOREM 4. *Under any odd prime p the following equalities are valid:*

$$\sum_{x=1}^{p} e^{2\pi i \frac{x^2}{p}} = \begin{cases} \sqrt{p} & \textit{if } \ p \equiv 1 \pmod 4, \\ i\sqrt{p} & \textit{if } \ p \equiv 3 \pmod 4. \end{cases}$$

Proof. Let us show at first that

$$\left| \sum_{\sqrt{p}<x<p} e^{2\pi i \frac{x^2}{4p}} \right| < \sqrt{p}. \tag{45}$$

Indeed, apply Abel's summation formula

$$\sum_{x=q+1}^{p-1} (u_x - u_{x-1})v_x = \sum_{x=q+1}^{p-1} u_x(v_x - v_{x+1}) + u_{p-1}v_p - u_q v_{q+1} \tag{46}$$

under $q = [\sqrt{p}]$ and

$$u_x = e^{2\pi i \frac{x(x+1)}{4p}}, \qquad v_x = \frac{1}{\sin \pi \frac{x}{2p}}.$$

Since, obviously,

$$u_{p-1}v_p = e^{2\pi i \frac{p-1}{4}} = (-1)^{\frac{p-1}{2}}$$

and

$$u_x - u_{x-1} = e^{2\pi i \frac{x^2}{4p}} \left(e^{2\pi i \frac{x}{4p}} - e^{-2\pi i \frac{x}{4p}} \right) = 2ie^{2\pi i \frac{x^2}{4p}} \sin \pi \frac{x}{2p},$$

then from (46) it follows that

$$2i \sum_{\sqrt{p}<x<p} e^{2\pi i \frac{x^2}{4p}} = \sum_{x=q+1}^{p-1} e^{2\pi i \frac{x(x+1)}{4p}} \left(\frac{1}{\sin \pi \frac{x}{2p}} - \frac{1}{\sin \pi \frac{x+1}{2p}} \right)$$

$$+ (-1)^{\frac{p-1}{2}} - \frac{e^{2\pi i \frac{q(q+1)}{4p}}}{\sin \pi \frac{q+1}{2p}}.$$

But then, observing that under $1 \leqslant x \leqslant p-1$

$$\left| \frac{1}{\sin \pi \frac{x}{2p}} - \frac{1}{\sin \pi \frac{x+1}{2p}} \right| = \frac{1}{\sin \pi \frac{x}{2p}} - \frac{1}{\sin \pi \frac{x+1}{2p}},$$

we get

$$2 \left| \sum_{\sqrt{p}<x<p} e^{2\pi i \frac{x^2}{4p}} \right| \leqslant \sum_{x=q+1}^{p-1} \left(\frac{1}{\sin \pi \frac{x}{2p}} - \frac{1}{\sin \pi \frac{x+1}{2p}} \right) + 1$$

$$+ \frac{1}{\sin \pi \frac{q+1}{2p}} = \frac{2}{\sin \pi \frac{q+1}{2p}}.$$

Since

$$\frac{2}{\sin \pi \frac{q+1}{2p}} \leqslant \frac{2p}{q+1} < 2\sqrt{p},$$

the estimate (45) follows.

Now, observing that

$$\operatorname{Re}(1-i) \sum_{1 \leqslant x < \sqrt{p}} e^{2\pi i \frac{x^2}{4p}} = \sum_{1 \leqslant x < \sqrt{p}} \left(\cos \pi \frac{x^2}{2p} + \sin \pi \frac{x^2}{2p} \right) > \sqrt{p} - 1,$$

and using the estimate (45), we get

$$\operatorname{Re}(1-i) \sum_{x=1}^{p-1} e^{2\pi i \frac{x^2}{4p}} \geqslant \operatorname{Re}(1-i) \sum_{1 \leqslant x < \sqrt{p}} e^{2\pi i \frac{x^2}{4p}} - \left| (1-i) \sum_{\sqrt{p}<x<p} e^{2\pi i \frac{x^2}{4p}} \right|$$

$$> \sqrt{p} - 1 - \sqrt{2}\sqrt{p} > -\sqrt{p}. \tag{47}$$

Let $p \equiv 1 \pmod 4$. Then by (44)

$$\sum_{x=1}^{p} e^{2\pi i \frac{x^2}{p}} = \pm\sqrt{p}, \tag{48}$$

i.e., this sum is a real number and, therefore,

$$\sum_{x=1}^{p} e^{2\pi i \frac{x^2}{p}} = 1 + \mathrm{Re} \sum_{x=1}^{p-1} e^{2\pi i \frac{x^2}{p}}.$$

Since by (35) under $p \equiv 1 \pmod 4$

$$\sum_{x=1}^{p-1} e^{2\pi i \frac{x^2}{p}} = (1-i) \sum_{x=1}^{p-1} e^{2\pi i \frac{x^2}{4p}},$$

then using the estimate (47) we get

$$\sum_{x=1}^{p} e^{2\pi i \frac{x^2}{p}} = 1 + \mathrm{Re}\,(1-i) \sum_{x=1}^{p-1} e^{2\pi i \frac{x^2}{4p}} > -\sqrt{p}.$$

By (48) the first assertion of the theorem is proved.

If $p \equiv 3 \pmod 4$, then by (35) and (44)

$$\frac{1}{i} \sum_{x=1}^{p} e^{2\pi i \frac{x^2}{p}} = \pm\sqrt{p}, \qquad \sum_{x=1}^{p-1} e^{2\pi i \frac{x^2}{p}} = (1+i) \sum_{x=1}^{p-1} e^{2\pi i \frac{x^2}{4p}}, \tag{49}$$

and as above we get

$$\frac{1}{i} \sum_{x=1}^{p} e^{2\pi i \frac{x^2}{p}} = \mathrm{Re}\, \frac{1}{i} \sum_{x=1}^{p-1} e^{2\pi i \frac{x^2}{p}} = \mathrm{Re}\, \frac{1+i}{i} \sum_{x=1}^{p-1} e^{2\pi i \frac{x^2}{4p}}$$
$$= \mathrm{Re}\,(1-i) \sum_{x=1}^{p-1} e^{2\pi i \frac{x^2}{4p}} > -\sqrt{p}.$$

By virtue of the first equality of (49), the theorem is proved in full.

Note that the assertion of Theorem 4 can be written by means of one equality without singling out the cases of $p \equiv 1 \pmod 4$ and $p \equiv 3 \pmod 4$:

$$\sum_{x=1}^{p} e^{2\pi i \frac{x^2}{p}} = i^{\left(\frac{p-1}{2}\right)^2} \sqrt{p}. \tag{50}$$

Hence by (43) under any $a \not\equiv 0 \pmod p$ we obtain

$$\sum_{x=1}^{p} e^{2\pi i \frac{ax^2}{p}} = i^{\left(\frac{p-1}{2}\right)^2} \left(\frac{a}{p}\right) \sqrt{p}. \tag{51}$$

The equality (50) was proved under the assumption that p is an odd prime. Let us show that the same equality is valid for Gaussian sums with an arbitrary odd denominator q:

$$\sum_{x=1}^{q} e^{2\pi i \frac{x^2}{q}} = i^{\left(\frac{q-1}{2}\right)^2} \sqrt{q}. \tag{52}$$

At first we shall consider sums of the form

$$S(a, p^\alpha) = \sum_{x=1}^{p^\alpha} e^{2\pi i \frac{ax^2}{p^\alpha}},$$

where α is a positive integer, p an odd prime, and a prime to p. Using the induction with respect to α, it is easy to show that

$$S(a, p^\alpha) = \left(\frac{a}{p}\right)^\alpha i^{\left(\frac{p^\alpha - 1}{2}\right)^2} p^{\frac{\alpha}{2}}. \tag{53}$$

Indeed, under $\alpha = 1$ this equality coincides with the equality (51). Under $\alpha = 2$ it takes the form $S(\alpha, p^2) = p$ and is obtained with the help of the summation variable replacement:

$$\sum_{x=1}^{p^2} e^{2\pi i \frac{ax^2}{p^2}} = \sum_{y=1}^{p} \sum_{z=0}^{p-1} e^{2\pi i \frac{a(y+pz)^2}{p^2}} = \sum_{y=1}^{p} e^{2\pi i \frac{ay^2}{p^2}} \sum_{z=0}^{p-1} e^{2\pi i \frac{2ayz}{p}}$$

$$= p \sum_{y=1}^{p} e^{2\pi i \frac{ay^2}{p}} \delta_p(2ay) = p.$$

Let the equality (53) be proved for a certain $\alpha \geqslant 2$ and all lesser values α. Let us prove it for $\alpha + 1$.

Obviously

$$S(a, p^{\alpha+1}) = \sum_{x=1}^{p^{\alpha+1}} e^{2\pi i \frac{ax^2}{p^{\alpha+1}}} = \sum_{y=1}^{p^\alpha} \sum_{z=0}^{p-1} e^{2\pi i \frac{a(y+p^\alpha z)^2}{p^{\alpha+1}}}$$

$$= \sum_{y=1}^{p^\alpha} e^{2\pi i \frac{ay^2}{p^{\alpha+1}}} \sum_{z=0}^{p-1} e^{2\pi i \frac{2ayz}{p}} = p \sum_{y=1}^{p^\alpha} e^{2\pi i \frac{ay^2}{p^{\alpha+1}}} \delta_p(2ay).$$

Observing that in the last sum the only summands with y multiple of p do not equal zero and that $p^2 \equiv 1 \pmod 8$, we obtain

$$S(a, p^{\alpha+1}) = p \sum_{y=1}^{p^{\alpha-1}} e^{2\pi i \frac{ay^2}{p^{\alpha-1}}} = pS(a, p^{\alpha-1})$$
$$= \left(\frac{a}{p}\right)^{\alpha-1} i^{\left(\frac{p^{\alpha-1}-1}{2}\right)^2} p^{1+\frac{\alpha-1}{2}}$$
$$= \left(\frac{a}{p}\right)^{\alpha+1} i^{\left(\frac{p^{\alpha+1}-1}{2}\right)^2} p^{\frac{\alpha+1}{2}}.$$

The equality (53) is proved in full.

Now let $q > 1$ be an arbitrary odd number. Write the prime factorization of q in the form $q = p_1^{\alpha_1} \dots p_s^{\alpha_s}$ and determine $a_1, \dots, a_s$ from the congruence

$$a_1 p_2^{\alpha_2} \dots p_s^{\alpha_s} + \dots + p_1^{\alpha_1} \dots p_{s-1}^{\alpha_{s-1}} a_s \equiv 1 \pmod{p_1^{\alpha_1} \dots p_s^{\alpha_s}}. \tag{54}$$

We shall assume that in the product $p_1^{\alpha_1} \dots p_s^{\alpha_s}$ the odd powers of the primes are put on the first r places. Since the equality (53) can be rewritten in the form

$$S(a, p^\alpha) = \begin{cases} \left(\frac{a}{p}\right) i^{\left(\frac{p-1}{2}\right)^2} p^{\frac{\alpha}{2}} & \text{if } \alpha \equiv 1 \pmod 2, \\ p^{\frac{\alpha}{2}} & \text{if } \alpha \equiv 0 \pmod 2, \end{cases}$$

then using the multiplication formula (34), we get

$$\sum_{x=1}^{q} e^{2\pi i \frac{x^2}{q}} = \prod_{\nu=1}^{s} \sum_{x_\nu=1}^{p_\nu^{\alpha_\nu}} e^{2\pi i \frac{a_\nu x_\nu^2}{p_\nu^{\alpha_\nu}}}$$
$$= \sqrt{q} \left(\frac{a_1}{p_1}\right) \dots \left(\frac{a_r}{p_r}\right) i^{\left(\frac{p_1-1}{2}\right)^2 + \dots + \left(\frac{p_r-1}{2}\right)^2}. \tag{55}$$

From the determination (54) it follows that

$$p_1^{\alpha_1} \dots a_\nu \dots p_s^{\alpha_s} \equiv 1 \pmod{p_\nu},$$

and since

$$\alpha_\nu \equiv \begin{cases} 1 \pmod 2 & \text{if } 1 \leqslant \nu \leqslant r, \\ 0 \pmod 2 & \text{if } r < \nu \leqslant s, \end{cases}$$

then, obviously, under $1 \leqslant \nu \leqslant r$

$$\left(\frac{p_1 \dots a_\nu \dots p_r}{p_\nu}\right) = 1, \qquad \left(\frac{a_\nu}{p_\nu}\right) = \left(\frac{p_1 \dots p_{\nu-1} p_{\nu+1} \dots p_r}{p_\nu}\right),$$

$$\left(\frac{a_1}{p_1}\right)\dots\left(\frac{a_r}{p_r}\right) = \prod_{\substack{j,k=1\\ j\neq k}}^{r} \left(\frac{p_j}{p_k}\right) = \prod_{\substack{j,k=1\\ j<k}}^{r} \left(\frac{p_j}{p_k}\right)\left(\frac{p_k}{p_j}\right). \tag{56}$$

Determine the quantities β_j and γ_r by the equalities

$$\beta_j = \frac{p_j - 1}{2}, \qquad \gamma_r = \sum_{\substack{j,k=1\\ j<k}}^{r} \beta_j\beta_k.$$

Then using the law of reciprocity of quadratic residues in the form

$$\left(\frac{p_j}{p_k}\right)\left(\frac{p_k}{p_j}\right) = (-1)^{\beta_j\beta_k} = i^{2\beta_j\beta_k} \qquad (j \neq k),$$

from (55) and (56) we obtain

$$\left(\frac{a_1}{p_1}\right)\dots\left(\frac{a_r}{p_r}\right) = \prod_{\substack{j,k=1\\ j<k}}^{r} i^{2\beta_j\beta_k} = i^{2\gamma_r},$$

$$\sum_{x=1}^{q} e^{2\pi i \frac{x^2}{q}} = i^{\beta_1^2+\dots+\beta_r^2+2\gamma_r}\sqrt{q} = i^{(\beta_1+\dots+\beta_r)^2}\sqrt{q}. \tag{57}$$

Since, obviously,

$$(\beta_1 + \dots + \beta_r)^2 = \left(\frac{p_1 - 1}{2} + \dots + \frac{p_r - 1}{2}\right)^2 \equiv \left(\frac{p_1\dots p_r - 1}{2}\right)^2$$
$$\equiv \left(\frac{p_1^{\alpha_1}\dots p_r^{\alpha_r} - 1}{2}\right)^2 \equiv \left(\frac{q-1}{2}\right)^2 \pmod 4,$$

then from (57) it follows, that for any odd q the equality (52) is valid:

$$\sum_{x=1}^{q} e^{2\pi i \frac{x^2}{q}} = i^{(\beta_1+\dots+\beta_r)^2}\sqrt{q} = i^{\left(\frac{q-1}{2}\right)^2}\sqrt{q}.$$

The exact value of Gaussian sums is known for an arbitrary even q, too. If $q \equiv 2 \pmod 4$, then according to Theorem 3 the Gaussian sum vanishes. Under $q \equiv 0 \pmod 4$ it can be shown that

$$\sum_{x=1}^{q} e^{2\pi i \frac{x^2}{q}} = (1+i)\sqrt{q}.$$

Thus the total description of the Gaussian sums magnitude is given by the equalities

$$\sum_{x=1}^{q} e^{2\pi i \frac{x^2}{q}} = \begin{cases} i^{\left(\frac{q-1}{2}\right)^2}\sqrt{q} & \text{if } q \equiv 1 \pmod 2,\\ (1+i)\sqrt{q} & \text{if } q \equiv 0 \pmod 4,\\ 0 & \text{if } q \equiv 2 \pmod 4. \end{cases} \tag{58}$$

§ 4. Simplest complete sums

An appropriate generalization of Gaussian sums is a complete rational sum of the form

$$S(a,q)=\sum_{x=1}^{q} e^{2\pi i\frac{ax^n}{q}}, \tag{59}$$

where a and q are coprime and $n\geqslant 2$. In distinction from the Gaussian sums ($n=2$), an explicit expression for the sums (59) under $n>2$ is not known, but for them it is easy to establish the estimates, whose order can not be improved further. It is quite easy to obtain the estimate

$$\left|\sum_{x=1}^{p} e^{2\pi i\frac{ax^n}{p}}\right|\leqslant n\sqrt{p}, \tag{60}$$

where p is a prime.

Indeed, let $T(b)$ and T be the number of solutions of the congruences $az^n\equiv b \pmod p$ and $x^n\equiv y^n \pmod p$, respectively. Using properties of binomial congruences, we get

$$T(b)\leqslant d,\qquad T=1+d(p-1), \tag{61}$$

where $d=(n,p-1)$. On the other hand, by Lemma 2

$$T=\sum_{x,y=1}^{p}\delta_p(x^n-y^n)=\frac{1}{p}\sum_{x,y,z=1}^{p} e^{2\pi i\frac{(x^n-y^n)z}{p}}$$

$$=p+\frac{1}{p}\sum_{z=1}^{p-1}\sum_{x,y=1}^{p} e^{2\pi i\frac{z(x^n-y^n)}{p}}=p+\frac{1}{p}\sum_{z=1}^{p-1}\left|\sum_{x=1}^{p} e^{2\pi i\frac{zx^n}{p}}\right|^2.$$

Therefore, by (61)

$$\sum_{z=1}^{p-1}\left|\sum_{x=1}^{p} e^{2\pi i\frac{zx^n}{p}}\right|^2=p(T-p)=(d-1)p(p-1). \tag{62}$$

Since by (25) under $1\leqslant z\leqslant p-1$

$$\left|\sum_{x=1}^{p} e^{2\pi i\frac{ax^n}{p}}\right|^2=\left|\sum_{x=1}^{p} e^{2\pi i\frac{a(zx)^n}{p}}\right|^2,$$

then carrying out the summation over z, we obtain

$$(p-1)\left|\sum_{x=1}^{p} e^{2\pi i\frac{ax^n}{p}}\right|^2=\sum_{z=1}^{p-1}\left|\sum_{x=1}^{p} e^{2\pi i\frac{az^nx^n}{p}}\right|^2.$$

Here we group the summands with $az^n \equiv b \pmod p$. Then, using the estimate (61) and the equality (62), we get

$$\left|\sum_{x=1}^{p} e^{2\pi i\frac{ax^n}{p}}\right|^2 = \frac{1}{p-1}\sum_{b=1}^{p-1} T(b)\left|\sum_{x=1}^{p} e^{2\pi i\frac{bx^n}{p}}\right|^2$$
$$\leqslant \frac{d}{p-1}\sum_{b=1}^{p-1}\left|\sum_{x=1}^{p} e^{2\pi i\frac{bx^n}{p}}\right|^2 = d(d-1)p.$$

Since $d \leqslant n$, the estimate (60) follows:

$$\left|\sum_{x=1}^{p} e^{2\pi i\frac{ax^n}{p}}\right| \leqslant \sqrt{d(d-1)p} < n\sqrt{p}.$$

The following theorem improves this estimate.

THEOREM 5. *Let $n \geqslant 2$, p be a prime, $(a,p)=1$, and $(n,p-1)=d$. Then*

$$\left|\sum_{x=1}^{p} e^{2\pi i\frac{ax^n}{p}}\right| \leqslant (d-1)\sqrt{p}. \tag{63}$$

Proof. At first we shall consider the case $d=n$. Using Lemma 2, we have

$$\sum_{\nu=1}^{p-1}\sum_{x=1}^{p} e^{2\pi i\frac{\nu x^n}{p}} = \sum_{x=1}^{p}\sum_{\nu=1}^{p-1} e^{2\pi i\frac{\nu x^n}{p}} = \sum_{x=1}^{p}\left[p\delta_p(x^n)-1\right] = 0. \tag{64}$$

Let g be a primitive root of p. Introduce the notation

$$S_\nu = \sum_{x=1}^{p} e^{2\pi i\frac{ag^{\nu-1}x^n}{p}}.$$

By (25) we obtain

$$S_{\nu+n} = \sum_{x=1}^{p} e^{2\pi i\frac{ag^{\nu-1}(gx)^n}{p}} = \sum_{x=1}^{p} e^{2\pi i\frac{ag^{\nu-1}x^n}{p}} = S_\nu$$

and, therefore,

$$\sum_{\nu=1}^{p-1} S_\nu = \frac{p-1}{n}(S_1+\ldots+S_n),$$
$$\sum_{\nu=1}^{p-1} |S_\nu|^2 = \frac{p-1}{n}(|S_1|^2+\ldots+|S_n|^2). \tag{65}$$

Observing that $ag^{\nu-1}$ and ν run through reduced residue systems modulo p simultaneously, by (62) and (64) we get

$$\sum_{\nu=1}^{p-1} S_\nu = \sum_{\nu=1}^{p-1}\sum_{x=1}^{p} e^{2\pi i \frac{ag^{\nu-1}x^n}{p}} = \sum_{\nu=1}^{p-1}\sum_{x=1}^{p} e^{2\pi i \frac{\nu x^n}{p}} = 0,$$

$$\sum_{\nu=1}^{p-1} |S_\nu|^2 = \sum_{\nu=1}^{p-1}\left|\sum_{x=1}^{p} e^{2\pi i \frac{ag^{\nu-1}x^n}{p}}\right|^2 = \sum_{\nu=1}^{p-1}\left|\sum_{x=1}^{p} e^{2\pi i \frac{\nu x^n}{p}}\right|^2 = (n-1)p(p-1).$$

Hence according to (65) it follows, that

$$S_1 + \ldots + S_n = 0 \quad \text{and} \quad |S_1|^2 + \ldots + |S_n|^2 = n(n-1)p.$$

But then

$$|S_1|^2 = |S_2 + \ldots + S_n|^2 \leqslant (n-1)\big(|S_2|^2 + \ldots + |S_n|^2\big),$$
$$|S_1|^2 = n(n-1)p - \big(|S_2|^2 + \ldots + |S_n|^2\big) \leqslant n(n-1)p - \frac{1}{n-1}|S_1|^2,$$

and, therefore,

$$|S_1|^2 \leqslant (n-1)^2 p. \tag{66}$$

Since by definition

$$S_1 = \sum_{x=1}^{p} e^{2\pi i \frac{ax^n}{p}},$$

then from (66) we get the theorem assertion for the case $(n, p-1) = n$

$$\left|\sum_{x=1}^{p} e^{2\pi i \frac{ax^n}{p}}\right| \leqslant (n-1)\sqrt{p}. \tag{67}$$

Now we shall consider the case $(n, p-1) = d$, where $d \leqslant n$. Denote the least non-negative residues of x^n and x^d modulo p by r_x and t_x, respectively. Observing that the quantities $r_1, \ldots, r_p$ form a permutation of the quantities $t_1, \ldots, t_p$ we obtain

$$\sum_{x=1}^{p} e^{2\pi i \frac{ar_x}{p}} = \sum_{x=1}^{p} e^{2\pi i \frac{at_x}{p}}$$

and, therefore,

$$\sum_{x=1}^{p} e^{2\pi i \frac{ax^n}{p}} = \sum_{x=1}^{p} e^{2\pi i \frac{ax^d}{p}}. \tag{68}$$

Since $(d, p-1) = d$, then by (67)

$$\left|\sum_{x=1}^{p} e^{2\pi i \frac{ax^d}{p}}\right| \leqslant (d-1)\sqrt{p}.$$

By (68) the theorem is proved in full.

Let's discuss the question about the possibility of further improvement of the estimate (63). Choose the quantity a in such a way that

$$\left|\sum_{x=1}^{p} e^{2\pi i \frac{ax^n}{p}}\right| = \max_{1 \leqslant \nu < p} \left|\sum_{x=1}^{p} e^{2\pi i \frac{\nu x^n}{p}}\right|.$$

Then using the equality (62), under $(n, p-1) = d$ we get

$$\left|\sum_{x=1}^{p} e^{2\pi i \frac{ax^n}{p}}\right|^2 = \max_{1 \leqslant \nu < p} \left|\sum_{x=1}^{p} e^{2\pi i \frac{\nu x^n}{p}}\right|^2$$
$$\geqslant \frac{1}{p-1} \sum_{\nu=1}^{p-1} \left|\sum_{x=1}^{p} e^{2\pi i \frac{\nu x^n}{p}}\right|^2 = (d-1)p.$$

Hence by Theorem 5 it follows that

$$\sqrt{d-1}\,\sqrt{p} \leqslant \left|\sum_{x=1}^{p} e^{2\pi i \frac{ax^n}{p}}\right| \leqslant (d-1)\sqrt{p}. \tag{69}$$

The inequalities (69) show that under fixed n and increasing p the estimate obtained in Theorem 5 has the order which can not be improved. Moreover, it can be shown [19], that in the estimate (63) it is impossible not only to improve the order, but to replace $(d-1)\sqrt{p}$ by the quantity $(1-\varepsilon)(d-1)\sqrt{p}$ under any $\varepsilon > 0$ either.

Now we shall consider sums with a prime-power denominator.

LEMMA 4. *Let* $2 \leqslant \alpha \leqslant n$, p *be a prime* $> n$, $(a_1, \ldots, a_n, p) = 1$, *and* $f(x) = a_1 x + \ldots + a_n x^n$. *Then the following estimate holds:*

$$\left|\sum_{x=1}^{p^\alpha} e^{2\pi i \frac{f(x)}{p^\alpha}}\right| \leqslant (n-1)p^{\alpha-1}.$$

Proof. Let y and z run through complete residue sets modulo $p^{\alpha-1}$ and p, respectively. Then the sum $y + p^{\alpha-1}z$ runs through a complete residue set modulo p^α and, since $\alpha \geqslant 2$ and $p > 2$, we have

$$f(y + p^{\alpha-1}z) \equiv f(y) + f'(y)p^{\alpha-1}z \pmod{p^\alpha}.$$

Therefore,

$$\sum_{x=1}^{p^\alpha} e^{2\pi i \frac{f(x)}{p^\alpha}} = \sum_{y=1}^{p^{\alpha-1}} \sum_{z=1}^{p} e^{2\pi i \frac{f(y+p^{\alpha-1}z)}{p^\alpha}}$$
$$= \sum_{y=1}^{p^{\alpha-1}} e^{2\pi i \frac{f(y)}{p^\alpha}} \sum_{z=1}^{p} e^{2\pi i \frac{f'(y)z}{p}}$$
$$= p \sum_{y=1}^{p^{\alpha-1}} e^{2\pi i \frac{f(y)}{p^\alpha}} \delta_p[f'(y)]. \tag{70}$$

But then

$$\left| \sum_{x=1}^{p^\alpha} e^{2\pi i \frac{f(x)}{p^\alpha}} \right| \leqslant p \sum_{y=1}^{p^{\alpha-1}} \delta_p[f'(y)] = p^{\alpha-1} \sum_{y=1}^{p} \delta_p[f'(y)] = p^{\alpha-1}T, \tag{71}$$

where T is the number of solutions of the congruence

$$f'(y) \equiv 0 \pmod{p}.$$

Since $(a_1, \ldots, a_n, p) = 1$ and p is a prime $> n$, then at least one of the coefficients of the polynomial $f'(y) = a_1 + 2a_2 y + \ldots + na_n y^{n-1}$ is prime to p and, therefore, $T \leqslant n-1$. Substituting this estimate into (71), we obtain the lemma assertion.

It is easy to improve this result for polynomials of a special form. Let p be prime, $(a,p) = 1$, and

$$S(a,p^\alpha) = \sum_{x=1}^{p^\alpha} e^{2\pi i \frac{ax^n}{p^\alpha}}.$$

Let's show that under $\alpha \geqslant 2$ and $n \geqslant 3$ the following equalities hold:

$$S(a,p^\alpha) = \begin{cases} p^{\alpha-1} & \text{if } 2 \leqslant \alpha \leqslant n \text{ and } (n,p)=1, \\ p^{n-1}S(a,p^{\alpha-n}) & \text{if } \alpha \geqslant n+1. \end{cases} \tag{72}$$

Indeed, from (70) it follows that

$$S(a,p^\alpha) = p \sum_{y=1}^{p^{\alpha-1}} e^{2\pi i \frac{ay^n}{p^\alpha}} \delta_p(nay^{n-1}).$$

Hence under $(n,p) = 1$ we get the first equality of (72):

$$S(a,p^\alpha) = p \sum_{y=1}^{p^{\alpha-1}} e^{2\pi i \frac{ay^n}{p^\alpha}} \delta_p(y) = p \sum_{y=1}^{p^{\alpha-2}} e^{2\pi i \frac{a(py)^n}{p^\alpha}} = p^{\alpha-1}.$$

Now let $\alpha \geqslant n+1$ and p^β be the greatest power of p, which divides n. Then, using the estimate $\alpha \geqslant p^\beta + 1 \geqslant (p-1)\beta + 2$ and considering separately the cases $\alpha \geqslant 2\beta + 2$ and $n+1 \leqslant \alpha \leqslant 2\beta + 1$ we obtain

$$(y + p^{\alpha-\beta-1}z)^n \equiv y^n + np^{\alpha-\beta-1}y^{n-1}z \pmod{p^\alpha}.$$

Therefore,

$$S(a, p^\alpha) = \sum_{y=1}^{p^{\alpha-\beta-1}} \sum_{z=1}^{p^{\beta+1}} e^{2\pi i \frac{a(y+p^{\alpha-\beta-1}z)^n}{p^\alpha}}$$

$$= \sum_{y=1}^{p^{\alpha-\beta-1}} e^{2\pi i \frac{ay^n}{p^\alpha}} \sum_{z=1}^{p^{\beta+1}} e^{2\pi i \frac{any^{n-1}z}{p^{\beta+1}}}$$

$$= p^{\beta+1} \sum_{y=1}^{p^{\alpha-\beta-1}} e^{2\pi i \frac{ay^n}{p^\alpha}} \delta_{p^{\beta+1}}(any^{n-1}).$$

Hence, since $(an, p^{\beta+1}) = p^\beta$, we have

$$S(a, p^\alpha) = p^{\beta+1} \sum_{y=1}^{p^{\alpha-\beta-1}} e^{2\pi i \frac{ay^n}{p^\alpha}} \delta_p(y)$$

$$= p^{\beta+1} \sum_{y=1}^{p^{\alpha-\beta-2}} e^{2\pi i \frac{ay^n}{p^{\alpha-n}}} = p^{n-1} S(a, p^{\alpha-n}).$$

Thus the assertion (72) is proved in full.

THEOREM 6. *Let n and q be arbitrary positive integers and $(a, q) = 1$. Then for the sum*

$$S(a, q) = \sum_{x=1}^{q} e^{2\pi i \frac{ax^n}{q}}$$

the estimate

$$|S(a, q)| \leqslant n^{n^6} q^{1-\frac{1}{n}} \tag{73}$$

holds.

Proof. Since $(a, q) = 1$, then under $n = 1$ and $n = 2$ the estimate (73) follows from Lemma 2 and Theorem 3, respectively:

$$\sum_{x=1}^{q} e^{2\pi i \frac{ax}{q}} = 0, \qquad \left| \sum_{x=1}^{q} e^{2\pi i \frac{ax^2}{q}} \right| \leqslant \sqrt{2q}.$$

Therefore it suffices to consider the case $n \geqslant 3$.

At first we shall show that for any prime p under $\alpha \geqslant 1$, $n \geqslant 3$, and $(a,p) = 1$

$$|S(a,p^\alpha)| \leqslant C_p(n) p^{\alpha\left(1-\frac{1}{n}\right)}, \tag{74}$$

where

$$C_p(n) = \begin{cases} n & \text{if } p < n^6, \\ 1 & \text{if } p > n^6. \end{cases}$$

Indeed, under $\alpha = 1$ by (60)

$$|S(a,p)| \leqslant n\sqrt{p} < \begin{cases} np^{1-\frac{1}{n}} & \text{if } p < n^6, \\ p^{1-\frac{1}{n}} & \text{if } p > n^6. \end{cases}$$

Let $2 \leqslant \alpha \leqslant n$ and $(n,p) = p$. Then $p \leqslant n$ and, using the trivial estimate, we get

$$|S(a,p^\alpha)| \leqslant p^\alpha \leqslant p^{\alpha\left(1-\frac{1}{n}\right)} p \leqslant np^{\alpha\left(1-\frac{1}{n}\right)}.$$

Let finally $2 \leqslant \alpha \leqslant n$ and $(n,p) = 1$. Then by (72)

$$|S(a,p^\alpha)| = p^{\alpha-1} = p^{\alpha\left(1-\frac{1}{\alpha}\right)} \leqslant p^{\alpha\left(1-\frac{1}{n}\right)}.$$

Thus the estimate (74) is satisfied under $1 \leqslant \alpha \leqslant n$. Apply the induction. Let under $1 + (k-1)n \leqslant \alpha \leqslant kn$ this estimate be valid for a certain $k \geqslant 1$.

We shall show that the estimate is valid under $1 + kn \leqslant \alpha \leqslant (k+1)n$ as well. Since it is plain that $1 + (k-1)n \leqslant \alpha - n \leqslant 1 + kn$ then using the equality (72), by the induction hypothesis we get

$$|S(a,p^\alpha)| = p^{n-1}|S(a,p^{\alpha-n})| \leqslant p^{n-1} C_p(n) p^{(\alpha-n)\left(1-\frac{1}{n}\right)} = C_p(n) p^{\alpha\left(1-\frac{1}{n}\right)}.$$

Thus the estimate (74) is proved for any $\alpha \geqslant 1$.

Let now $q = p_1^{\alpha_1} \dots p_s^{\alpha_s}$ be the prime factorization of q. Using the multiplication formula (34), we obtain

$$\sum_{x=1}^{q} e^{2\pi i \frac{ax^n}{q}} = \prod_{\nu=1}^{s} \sum_{x_\nu=1}^{p_\nu^{\alpha_\nu}} e^{2\pi i \frac{ab_\nu x^n}{p_\nu^{\alpha_\nu}}}, \tag{75}$$

where the quantities b_ν are coprime with p_ν. We determine a_ν with the help of the equalities

$$a_\nu = ab_\nu \qquad (\nu = 1, 2, \dots, s).$$

Then, obviously, $(a_\nu, p_\nu) = 1$ and by (75)

$$S(a,q) = S(a_1, p_1^{\alpha_1}) \dots S(a_s, p_s^{\alpha_s}).$$

Hence, using the estimate (74) and observing that the number of primes less than n^6 does not exceed n^6, we get the theorem assertion:

$$|S(a,q)| \leqslant C_{p_1}(n)p_1^{\alpha_1\left(1-\frac{1}{n}\right)} \dots C_{p_s}(n)p_s^{\alpha_s\left(1-\frac{1}{n}\right)} \leqslant n^{n^6} q^{1-\frac{1}{n}}.$$

Note that under $n > 2$, $q = p^n$ and $(a,p) = 1$ by (72) for any prime p

$$\sum_{x=1}^{p^n} e^{2\pi i \frac{ax^n}{p^n}} = p^{n-1} = p^{n\left(1-\frac{1}{n}\right)}.$$

Therefore, in this case

$$S(a,q) = q^{1-\frac{1}{n}}.$$

Thus under fixed n and increasing q the order of the estimate (73) can not be improved.

Let $f(x) = a_1 x + \dots + a_n x^n$, $(a_1, \dots, a_n, q) = 1$ and $S(q)$ be a complete rational exponential sum of the general form

$$S(q) = \sum_{x=1}^{q} e^{2\pi i \frac{f(x)}{q}}. \tag{76}$$

In Theorem 6 the estimate

$$|S(q)| \leqslant C(n) q^{1-\frac{1}{n}}, \tag{77}$$

where $C(n) = n^{n^6}$, was proved for polynomials of the special form $f(x) = a_n x^n$. With the help of the significant complication of the proof technique, Hua Loo-Keng showed that under certain $C(n)$ the estimate (77) is valid for arbitrary complete rational sums (77) as well. A proof of an estimate close to (77) can be found in [16] and [44].

§ 5. Mordell's method

Let us consider a complete exponential sum with a prime denominator

$$S(p) = \sum_{x=1}^{p} e^{2\pi i \frac{a_1 x + \dots + a_n x^n}{p}}.$$

Mordell [36] proposed a method of such sums estimation based on the use of properties of the system of congruences

$$\left.\begin{array}{l} x_1 + \dots + x_n \equiv y_1 + \dots + y_n \\ \dots\dots\dots\dots\dots\dots\dots\dots \\ x_1^n + \dots + x_n^n \equiv y_1^n + \dots + y_n^n \end{array}\right\} \pmod p, \tag{78}$$

where p is a prime greater than n and the variables $x_1, \ldots, y_n$ run through complete residue sets modulo p independently.

First of all we shall prove a lemma about the number of solutions of a congruence system of a more general form.

LEMMA 5. *Let $q_1, \ldots, q_n$ be arbitrary positive integers, $q = \mathrm{LCM}(q_1, \ldots, q_n)$ and T_k be the number of solutions of the system of congruences*

$$\left.\begin{array}{l} x_1 + \ldots + x_k \equiv y_1 + \ldots + y_k \pmod{q_1} \\ \dots\dots\dots\dots\dots\dots\dots\dots \\ x_1^n + \ldots + x_k^n \equiv y_1^n + \ldots + y_k^n \pmod{q_n} \end{array}\right\}, \qquad 1 \leqslant x_\nu, y_\nu \leqslant q. \tag{79}$$

Then

$$T_k = \frac{1}{q_1 \cdots q_n} \sum_{a_1=1}^{q_1} \cdots \sum_{a_n=1}^{q_n} \left| \sum_{x=1}^{q} e^{2\pi i \left(\frac{a_1 x}{q_1} + \ldots + \frac{a_n x^n}{q_n} \right)} \right|^{2k}.$$

Proof. Since the product

$$\delta_{q_1}(x_1 + \ldots - y_k) \ldots \delta_{q_n}(x_1^n + \ldots - y_k^n)$$

equals unity, if numbers $x_1, \ldots, y_k$ satisfy the congruence system (79), and vanishes otherwise, then, obviously,

$$T_k = \sum_{x_1, \ldots, y_k = 1}^{q} \delta_{q_1}(x_1 + \ldots - y_k) \ldots \delta_{q_n}(x_1^n + \ldots - y_k^n).$$

Hence, using Lemma 2, we get the assertion of Lemma 5:

$$\begin{aligned} T_k &= \frac{1}{q_1 \cdots q_n} \sum_{\substack{a_1, \ldots, a_n \\ 1 \leqslant a_\nu \leqslant q_\nu}} \sum_{x_1, \ldots, y_k = 1}^{q} e^{2\pi i \left(\frac{a_1 (x_1 + \ldots - y_k)}{q_1} + \ldots + \frac{a_n (x_1^n + \ldots - y_k^n)}{q_n} \right)} \\ &= \frac{1}{q_1 \cdots q_n} \sum_{a_1=1}^{q_1} \cdots \sum_{a_n=1}^{q_n} \left| \sum_{x=1}^{q} e^{2\pi i \left(\frac{a_1 x}{q_1} + \ldots + \frac{a_n x^n}{q_n} \right)} \right|^{2k}. \end{aligned}$$

In particular, under $k = n$ and $q_1 = \ldots = q_n = p$ it follows from Lemma 5, that

$$T_n = \frac{1}{p^n} \sum_{a_1, \ldots, a_n = 1}^{p} \left| \sum_{x=1}^{p} e^{2\pi i \frac{a_1 x + \ldots + a_n x^n}{p}} \right|^{2n}, \tag{80}$$

where T_n is the number of solutions of the system (78).

LEMMA 6. *Under any $n \geqslant 1$ and a prime $p > n$, the number of solutions of the system (78) satisfies inequality $T_n \leqslant n!\, p^n$.*

Proof. Let $\lambda_1, \ldots, \lambda_n$ be fixed integers, $0 \leqslant \lambda_\nu \leqslant p-1$, and let $T(\lambda_1, \ldots, \lambda_n)$ be the number of solutions of the system of congruences

$$\left.\begin{array}{l} x_1 + \ldots + x_n \equiv \lambda_1 \\ \ldots\ldots\ldots\ldots\ldots\ldots \\ x_1^n + \ldots + x_n^n \equiv \lambda_n \end{array}\right\} \pmod p \qquad 1 \leqslant x_\nu \leqslant p. \tag{81}$$

We shall show that

$$T(\lambda_1, \ldots, \lambda_n) \leqslant n!. \tag{82}$$

Indeed, we introduce the following notation for the elementary symmetric functions and the sums of powers of quantities $x_1, \ldots, x_n$:

$$\sigma_1 = x_1 + \ldots + x_n, \ldots, \sigma_n = x_1 \ldots x_n,$$
$$s_1 = x_1 + \ldots + x_n, \ldots, s_n = x_1^n + \ldots + x_n^n.$$

Let $x_1, \ldots, x_n$ be an arbitrary solution of the system (81). Then, obviously,

$$s_1 \equiv \lambda_1, \ldots, s_n \equiv \lambda_n \pmod p,$$

and using the Newton recurrence formula

$$\nu\sigma_\nu = s_1\sigma_{\nu-1} - s_2\sigma_{\nu-2} + \ldots \mp s_{\nu-1}\sigma_1 \pm s_\nu,$$

under $\nu = 1, 2, \ldots, n$ we have

$$\nu\sigma_\nu \equiv \lambda_1\sigma_{\nu-1} - \lambda_2\sigma_{\nu-2} + \ldots \mp \lambda_{\nu-1}\sigma_1 \pm \lambda_\nu \pmod p. \tag{83}$$

Since p is a prime greater than n, then $(\nu, p) = 1$ and the congruence (83) is soluble for σ_ν. From (83) we get successively

$$\sigma_1 \equiv \mu_1, \ldots, \sigma_n \equiv \mu_n \pmod p \qquad (0 \leqslant \mu_\nu \leqslant p-1),$$

where the values $\mu_1, \ldots, \mu_n$ are determined uniquely by setting quantities $\lambda_1, \ldots, \lambda_n$. But then every solution of the system (81) coincides with one of the permutations of the roots of the congruence

$$x^n - \mu_1 x^{n-1} + \ldots \pm \mu_n \equiv 0 \pmod p$$

with fixed coefficients and, therefore,

$$T(\lambda_1, \ldots, \lambda_n) \leqslant n!.$$

Now, since

$$T_n = \sum_{y_1, \ldots, y_n = 1}^{p} T(y_1 + \ldots + y_n, \ldots, y_1^n + \ldots + y_n^n),$$

we get the lemma assertion:

$$T_n \leqslant \sum_{y_1, \ldots, y_n = 1}^{p} n! = n!\,p^n.$$

Note. From this lemma and the equality (80) it follows immediately, that under any $n \geqslant 1$ and a prime $p > n$ the following estimate holds:

$$\sum_{a_1,\ldots,a_n=1}^{p} \left| \sum_{x=1}^{p} e^{2\pi i \frac{a_1x+\ldots+a_nx^n}{p}} \right|^{2n} = p^n T_n \leqslant n!\, p^{2n}.$$

THEOREM 7. *Let $n \geqslant 2$, p be a prime greater than n, $(a_1, \ldots, a_n, p) = 1$ and $f(x) = a_1x + \ldots + a_nx^n$. Then*

$$\left| \sum_{x=1}^{p} e^{2\pi i \frac{f(x)}{p}} \right| \leqslant np^{1-\frac{1}{n}}.$$

Proof. At first we shall consider the case $(a_n, p) = 1$. Let integers λ and μ vary in the bounds $1 \leqslant \lambda \leqslant p-1$, $1 \leqslant \mu \leqslant p$. Arrange the polynomial $f(\lambda x + \mu)$ in the ascending order of powers of x

$$f(\lambda x + \mu) = b_0(\lambda, \mu) + b_1(\lambda, \mu)x + \ldots + b_n(\lambda, \mu)x^n \tag{84}$$

and observe that

$$b_n(\lambda, \mu) = a_n\lambda^n \qquad \text{and} \qquad b_{n-1}(\lambda, \mu) = (na_n\mu + a_{n-1})\lambda^{n-1}.$$

Denote the number of solutions of the system

$$\left.\begin{array}{l} b_1(\lambda, \mu) \equiv b_1 \\ \ldots\ldots\ldots\ldots\ldots\ldots \\ b_n(\lambda, \mu) \equiv b_n \end{array}\right\} \pmod{p} \tag{85}$$

by $H(b_1, \ldots, b_n)$. It is plain, that $H(b_1, \ldots, b_n)$ does not exceed the number of solutions of the system made up of the last two congruences of the system (85):

$$\left.\begin{array}{r} (na_n\mu + a_{n-1})\lambda^{n-1} \equiv b_{n-1} \\ a_n\lambda^n \equiv b_n \end{array}\right\} \pmod{p},$$

and, therefore, since $(na_n, p) = 1$ and $(\lambda, p) = 1$,

$$H(b_1, \ldots, b_n) \leqslant n. \tag{86}$$

By (25) for complete sums the equality

$$\left| \sum_{x=1}^{p} e^{2\pi i \frac{f(x)}{p}} \right|^{2n} = \left| \sum_{x=1}^{p} e^{2\pi i \frac{f(\lambda x+\mu)}{p}} \right|^{2n}$$

holds. Hence by (84) after the summation with respect to λ and μ, we have

$$\left|\sum_{x=1}^{p} e^{2\pi i \frac{f(x)}{p}}\right|^{2n} = \frac{1}{p(p-1)} \sum_{\lambda=1}^{p-1} \sum_{\mu=1}^{p} \left|\sum_{x=1}^{p} e^{2\pi i \frac{f(\lambda x+\mu)}{p}}\right|^{2n}$$
$$= \frac{1}{p(p-1)} \sum_{\lambda=1}^{p-1} \sum_{\mu=1}^{p} \left|\sum_{x=1}^{p} e^{2\pi i \frac{b_1(\lambda,\mu)x+\ldots+b_n(\lambda,\mu)x^n}{p}}\right|^{2n}.$$

Grouping the summands with fixed values $b_1(\lambda, \mu), \ldots, b_n(\lambda, \mu)$ and using the estimate (86), we get

$$\left|\sum_{x=1}^{p} e^{2\pi i \frac{f(x)}{p}}\right|^{2n} = \frac{1}{p(p-1)} \sum_{b_1,\ldots,b_n=1}^{p} H(b_1, \ldots, b_n) \left|\sum_{x=1}^{p} e^{2\pi i \frac{b_1 x+\ldots+b_n x^n}{p}}\right|^{2n}$$
$$\leqslant \frac{n}{p(p-1)} \sum_{b_1,\ldots,b_n=1}^{p} \left|\sum_{x=1}^{p} e^{2\pi i \frac{b_1 x+\ldots+b_n x^n}{p}}\right|^{2n}.$$

Hence by the note of Lemma 6 we obtain the theorem assertion for the case $(a_n, p) = 1$:

$$\left|\sum_{x=1}^{p} e^{2\pi i \frac{f(x)}{p}}\right|^{2n} \leqslant \frac{n}{p(p-1)} n!\, p^{2n} < n^{2n} p^{2n-2},$$
$$\left|\sum_{x=1}^{p} e^{2\pi i \frac{f(x)}{p}}\right| < np^{1-\frac{1}{n}}.$$

Now we show that the general case $(a_1, \ldots, a_n, p) = 1$ can be reduced to the case when leading coefficient of the polynomial is prime to p.

Indeed, let $(a_s, p) = 1$ and $a_{s+1} \equiv \ldots \equiv a_n \equiv 0 \pmod p$, $1 \leqslant s \leqslant n$. Then we obtain

$$\left|\sum_{x=1}^{p} e^{2\pi i \frac{f(x)}{p}}\right| = \left|\sum_{x=1}^{p} e^{2\pi i \frac{a_1 x+\ldots+a_s x^s}{p}}\right| \leqslant sp^{1-\frac{1}{s}} \leqslant np^{1-\frac{1}{n}}.$$

The theorem is proved in full.

Note. A substantial improvement of Mordell's estimate was obtained by A. Weil [48], who showed, that under prime $p > n$ and $(a_n, \ldots, a_n, p) = 1$ the estimate

$$\left|\sum_{x=1}^{p} e^{2\pi i \frac{a_1 x+\ldots+a_n x^n}{p}}\right| \leqslant (n-1)\sqrt{p}$$

is valid.

§ 6. Systems of congruences

One of the main points of Mordell's method (§ 5) is the use of the estimation for the number of solutions of the congruence system

$$\left.\begin{array}{l} x_1+\ldots+x_n\equiv y_1+\ldots+y_n \\ \ldots\ldots\ldots\ldots\ldots\ldots\ldots \\ x_1^n+\ldots+x_n^n\equiv y_1^n+\ldots+y_n^n \end{array}\right\} \pmod p,$$

where p is a prime greater than n. Hereafter, congruences of the same form but with respect to distinct moduli being equal to growing powers of a prime p will be of great importance. For the first time such systems of congruences were applied by Yu. V. Linnik [34] for the estimation of Weyl's sums by Vinogradov's method.

LEMMA 7. *Let $n\geqslant 1$, $k\geqslant \frac{n(n+1)}{4}$, p be a prime greater than n, and let $T_k(p^n)$ be the number of solutions of the system of congruences*

$$\left.\begin{array}{ll} x_1+\ldots+x_k\equiv y_1+\ldots+y_k & \pmod p \\ \ldots\ldots\ldots\ldots\ldots\ldots\ldots\ldots\ldots & \\ x_1^n+\ldots+x_k^n\equiv y_1^n+\ldots+y_k^n & \pmod{p^n} \end{array}\right\}, \qquad 1\leqslant x_\nu, y_\nu\leqslant p^n. \tag{87}$$

Then

$$T_k(p^n)\leqslant n^{2k}p^{2nk-\frac{n(n+1)}{2}}.$$

Proof. Under $n=1$ the lemma assertion is evident, so it suffices to consider the case $n\geqslant 2$. Take $q_1=p,\ldots,q_n=p^n$ in Lemma 5.

Then we obtain

$$T_k(p^n)=p^{-\frac{n(n+1)}{2}}\sum_{a_1=1}^{p}\cdots\sum_{a_n=1}^{p^n}\left|\sum_{x=1}^{p^n}e^{2\pi i\left(\frac{a_1x}{p}+\ldots+\frac{a_nx^n}{p^n}\right)}\right|^{2k}.$$

We split up the domain of the summation over $a_1,\ldots,a_n$ into two parts:

$$\sum_{a_1=1}^{p}\cdots\sum_{a_n=1}^{p^n}=\sum_{a_1,\ldots,a_n}{}_1+\sum_{a_1,\ldots,a_n}{}_2,$$

where the summation in $\sum_1$ is extended over n-tuples $a_1\ldots a_n$ which satisfy

$$p\backslash a_2,\ p^2\backslash a_3,\ldots,p^{n-1}\backslash a_n,$$

and for $\sum_2$ only those n-tuples are taken into account, for which at least for one of ν in the interval $2\leqslant\nu\leqslant n$ $p^{\nu-1}$ is not a divisor of a_ν.

In the first case, determining $b_1, \ldots, b_n$ with the help of the equalities

$$a_1 = b_1,\ a_2 = b_2 p,\ \ldots,\ a_n = b_n p^{n-1},$$

we get

$$\sum_{x=1}^{p^n} e^{2\pi i\left(\frac{a_1 x}{p}+\ldots+\frac{a_n x^n}{p^n}\right)} = \sum_{x=1}^{p^n} e^{2\pi i\,\frac{b_1 x+\ldots+b_n x^n}{p}} = p^{n-1}\sum_{x=1}^{p} e^{2\pi i\,\frac{b_1 x+\cdots+b_n x^n}{p}}.$$

Therefore

$$\left|\sum_{x=1}^{p^n} e^{2\pi i\left(\frac{a_1 x}{p}+\ldots+\frac{a_n x^n}{p^n}\right)}\right|^{2k} = p^{2nk-2k}\left|\sum_{x=1}^{p} e^{2\pi i\,\frac{b_1 x+\ldots+b_n x^n}{p}}\right|^{2k} \leqslant p^{2nk-2n}\left|\sum_{x=1}^{p} e^{2\pi i\,\frac{b_1 x+\ldots+b_n x^n}{p}}\right|^{2n},$$

and using the note of Lemma 6, we obtain

$$\sum_{a_1,\ldots,a_n}{}_1 \left|\sum_{x=1}^{p^n} e^{2\pi i\left(\frac{a_1 x}{p}+\ldots+\frac{a_n x^n}{p^n}\right)}\right|^{2k} \leqslant p^{2nk-2n}\sum_{b_1,\ldots,b_n=1}^{p}\left|\sum_{x=1}^{p} e^{2\pi i\,\frac{b_1 x+\ldots+b_n x^n}{p}}\right|^{2n} \leqslant n!\,p^{2nk}. \tag{88}$$

In the second case, there exists an integer ν in the interval $2 \leqslant \nu \leqslant n$ such that $p^{n-1} \nmid a_\nu p^{n-\nu}$. Therefore,

$$(a_1 p^{n-1}, a_2 p^{n-2}, \ldots, a_n, p^n) = p^{n-\alpha},$$

where $2 \leqslant \alpha \leqslant n$. But then

$$a_1 p^{n-1} = b_1 p^{n-\alpha}, \ldots, a_n = b_n p^{n-\alpha}, \qquad (b_1, \ldots, b_n, p) = 1,$$

and using Lemma 4, we get

$$\left|\sum_{x=1}^{p^n} e^{2\pi i\left(\frac{a_1 x}{p}+\ldots+\frac{a_n x^n}{p^n}\right)}\right| = \left|\sum_{x=1}^{p^n} e^{2\pi i\,\frac{b_1 x+\ldots+b_n x^n}{p^\alpha}}\right| = p^{n-\alpha}\left|\sum_{x=1}^{p^\alpha} e^{2\pi i\,\frac{b_1 x+\ldots+b_n x^n}{p^\alpha}}\right| \leqslant (n-1)p^{n-1}.$$

Hence, since $k \geqslant \frac{n(n+1)}{4}$, it follows that

$$\sum_{a_1,\ldots,a_n}{}_2 \left| \sum_{x=1}^{p^n} e^{2\pi i \left(\frac{a_1 x}{p} + \ldots + \frac{a_n x^n}{p^n} \right)} \right|^{2k} \leqslant (n-1)^{2k} p^{2nk - 2k + \frac{n(n+1)}{2}} \leqslant (n-1)^{2k} p^{2nk}. \tag{89}$$

Now, observing that under $n \geqslant 2$

$$n^{2k} - (n-1)^{2k} \geqslant 2k(n-1)^{2k-1} > n(n-1)^{n-1} \geqslant n!$$

from (88) and (89) we obtain the lemma assertion:

$$\begin{aligned} T_k(p^n) &= p^{-\frac{n(n+1)}{2}} \Big(\sum_{a_1,\ldots,a_n}{}_1 + \sum_{a_1,\ldots,a_n}{}_2 \Big) \\ &\leqslant n!\, p^{2nk - \frac{n(n+1)}{2}} + (n-1)^{2k} p^{2nk - \frac{n(n+1)}{2}} \\ &\leqslant n^{2k} p^{2nk - \frac{n(n+1)}{2}}. \end{aligned}$$

Note. Let $T_k(P)$ denote the number of solutions of the system (87), when the domain of variables variation has the form

$$1 \leqslant x_j \leqslant P, \qquad 1 \leqslant y_j \leqslant P \qquad (j = 1, 2, \ldots, k).$$

If m is a positive integer, then under $P = mp^n$

$$T_k(mp^n) \leqslant (mn)^{2k} p^{2nk - \frac{n(n+1)}{2}}. \tag{90}$$

Indeed, using the complete sums property (26), we get

$$\sum_{x=1}^{mp^n} e^{2\pi i \left(\frac{a_1 x}{p} + \ldots + \frac{a_n x^n}{p^n} \right)} = m \sum_{x=1}^{p^n} e^{2\pi i \left(\frac{a_1 x}{p} + \ldots + \frac{a_n x^n}{p^n} \right)},$$

and, therefore,

$$\begin{aligned} T_k(mp^n) &= p^{-\frac{n(n+1)}{2}} \sum_{a_1=1}^{p} \cdots \sum_{a_n=1}^{p^n} \left| \sum_{x=1}^{mp^n} e^{2\pi i \left(\frac{a_1 x}{p} + \ldots + \frac{a_n x^n}{p^n} \right)} \right|^{2k} \\ &= m^{2k} p^{-\frac{n(n+1)}{2}} \sum_{a_1=1}^{p} \cdots \sum_{a_n=1}^{p^n} \left| \sum_{x=1}^{p^n} e^{2\pi i \left(\frac{a_1 x}{p} + \ldots + \frac{a_n x^n}{p^n} \right)} \right|^{2k} \\ &= m^{2k} T_k(p^n). \end{aligned}$$

Hence, since by Lemma 7

$$T_k(p^n) \leqslant n^{2k} p^{2nk - \frac{n(n+1)}{2}},$$

we obtain the estimate (90).

Let $\sum_{x_1,\ldots,x_n}^{p^\alpha}$ denote the sum, in which the summation variables $x_1, \ldots, x_n$ run through complete residue sets modulo p^α and belong to different classes modulo p.

LEMMA 8. *Let p be a prime greater than n, $\alpha \geqslant 2$, and $f(x) = a_1 x + \ldots + a_n x^n$. Let $S_\alpha(a_1, \ldots, a_n)$ be defined with the help of the equality*

$$S_\alpha(a_1, \ldots, a_n) = \sum_{x_1,\ldots,x_n}^{p^\alpha} e^{2\pi i \frac{f(x_1)+\ldots+f(x_n)}{p^\alpha}}.$$

Then

$$S_\alpha(a_1, \ldots, a_n) = \begin{cases} p^{(\alpha-1)n} S_1(b_1, \ldots, b_n) & \text{if} \quad a_\nu = p^{\alpha-1} b_\nu \quad (\nu = 1, 2, \ldots, n), \\ 0 & \text{otherwise.} \end{cases}$$

Proof. Let us change the variables

$$x_\nu = y_\nu + p^{\alpha-1} z_\nu \qquad (\nu = 1, 2, \ldots, n).$$

Since by the assumption the quantities $x_1, \ldots, x_n$ belong to different classes, then the quantities $y_1, \ldots, y_n$ belong to different classes modulo p as well. Therefore, using that

$$f(y_\nu + p^{\alpha-1} z_\nu) \equiv f(y_\nu) + f'(y_\nu) p^{\alpha-1} z_\nu \pmod{p^\alpha},$$

we obtain

$$\begin{aligned} S_\alpha(a_1, \ldots, a_n) &= \sum_{y_1,\ldots,y_n}^{p^{\alpha-1}} e^{2\pi i \frac{f(y_1)+\ldots+f(y_n)}{p^\alpha}} \sum_{z_1,\ldots,z_n=1}^{p} e^{2\pi i \frac{f'(y_1)z_1+\ldots+f'(y_n)z_n}{p}} \\ &= p^n \sum_{y_1,\ldots,y_n}^{p^{\alpha-1}} e^{2\pi i \frac{f(y_1)+\ldots+f(y_n)}{p^\alpha}} \delta_p\big[f'(y_1)\big] \ldots \delta_p\big[f'(y_n)\big]. \end{aligned} \tag{91}$$

Since $f'(y) = a_1 + 2a_2 y + \ldots + n a_n y^{n-1}$, then under prime $p > n$ and $(a_1, \ldots, a_n, p) = 1$ the congruence $f'(y) \equiv 0 \pmod{p}$ can be satisfied by at most $n - 1$ values of y from different classes modulo p. In the sum (91) the quantities $y_1, \ldots, y_n$ belong to different classes and, therefore,

$$\delta_p\big[f'(y_1)\big] \ldots \delta_p\big[f'(y_n)\big] = \begin{cases} 1 & \text{if} \quad (a_1, \ldots, a_n, p) = p, \\ 0 & \text{if} \quad (a_1, \ldots, a_n, p) = 1. \end{cases}$$

But then by (91) the sum $S_\alpha(a_1,\ldots,a_n)$ vanishes under $(a_1,\ldots,a_n,p) = 1$. If $(a_1,\ldots,a_n,p) = p$, then

$$S_\alpha(a_1,\ldots,a_n) = p^n \sum_{y_1,\ldots,y_n}^{p^{\alpha-1}} e^{2\pi i \frac{p^{-1}f(y_1)+\ldots+p^{-1}f(y_n)}{p^{\alpha-1}}}.$$

Thus

$$S_\alpha(a_1,\ldots,a_n) = \begin{cases} p^n S_{\alpha-1}(a_1p^{-1},\ldots,a_np^{-1}) & \text{if } (a_1,\ldots,a_n,p) = p \\ 0 & \text{if } (a_1,\ldots,a_n,p) = 1. \end{cases} \tag{92}$$

Applying the equality (92) to $S_{\alpha-1}(a_1p^{-1},\ldots,a_np^{-1})$ we get

$$S_\alpha(a_1,\ldots,a_n) = \begin{cases} p^{2n} S_{\alpha-2}(a_1p^{-2},\ldots,a_np^{-2}) & \text{if } (a_1,\ldots,a_n,p^2) = p^2, \\ 0 & \text{otherwise.} \end{cases}$$

Continue this process. Then after $\alpha - 1$ step we obtain the lemma assertion:

$$S_\alpha(a_1,\ldots,a_n) = \begin{cases} p^{(\alpha-1)n} S_1\left(a_1p^{-(\alpha-1)},\ldots,a_np^{-(\alpha-1)}\right) & \text{if } (a_1,\ldots,a_n,p^{\alpha-1}) = p^{\alpha-1}, \\ 0 & \text{otherwise.} \end{cases}$$

LEMMA 9. (Linnik's lemma). *Let $\lambda_1,\ldots,\lambda_n$ be fixed integers, p a prime greater than n and let $T^*(\lambda_1,\ldots,\lambda_n)$ be the number of solutions of the system of congruences*

$$\left.\begin{array}{l} x_1 + \ldots + x_n \equiv \lambda_1 \pmod p \\ \ldots\ldots\ldots\ldots\ldots\ldots\ldots \\ x_1^n + \ldots + x_n^n \equiv \lambda_n \pmod{p^n} \end{array}\right\},$$

where the variables run through complete residue sets modulo p^n and belong to different classes modulo p. Then

$$T^*(\lambda_1,\ldots,\lambda_n) \leqslant n!\, p^{\frac{n(n-1)}{2}}.$$

Proof. Let $f(x) = a_1p^{n-1}x + a_2p^{n-2}x^2 + \ldots + a_nx^n$ and according to the notation of Lemma 8

$$S_n(a_1p^{n-1},\ldots,a_n) = \sum_{x_1,\ldots,x_n}^{p^n} e^{2\pi i \frac{f(x_1)+\ldots+f(x_n)}{p^n}}.$$

Using Lemma 2 we obtain

$$T^*(\lambda_1,\ldots,\lambda_n) = \sum_{x_1,\ldots,x_n}^{p^n} \prod_{\nu=1}^{n} \delta_{p^\nu}(x_1^\nu + \ldots + x_n^\nu - \lambda_\nu),$$

$$p^{\frac{n(n+1)}{2}} T^*(\lambda_1,\ldots,\lambda_n) = \sum_{a_1,\ldots,a_n} \sum_{x_1,\ldots,x_n}^{p^n} e^{2\pi i \left(a_1 \frac{x_1+\ldots+x_n-\lambda_1}{p} + \ldots + a_n \frac{x_1^n+\ldots+x_n^n-\lambda_n}{p^n}\right)},$$

where the summation with respect to $a_1, \dots, a_n$ is extended over the domain $1 \leqslant a_1 \leqslant p, \dots, 1 \leqslant a_n \leqslant p^n$. Hence observing that

$$\sum_{x_1,\dots,x_n}^{p^n} e^{2\pi i\left(a_1 \frac{x_1+\dots+x_n}{p}+\dots+a_n\frac{x_1^n+\dots+x_n^n}{p^n}\right)}$$
$$= \sum_{x_1,\dots,x_n}^{p^n} e^{2\pi i \frac{f(x_1)+\dots+f(x_n)}{p^n}} = S_n(a_1p^{n-1}, \dots, a_n),$$

we have

$$T^*(\lambda_1, , \dots, \lambda_n)$$
$$= p^{-\frac{n(n+1)}{2}} \sum_{a_1=1}^{p} \dots \sum_{a_n=1}^{p^n} S_n(a_1p^{n-1}, \dots, a_n) e^{-2\pi i\left(\frac{a_1\lambda_1}{p}+\dots+\frac{a_n\lambda_n}{p^n}\right)}.$$

Determine the quantities $b_1, \dots, b_n$ with the help of the equalities

$$a_1 = b_1, \ a_2 = pb_2, \dots, a_n = p^{n-1}b_n.$$

According to Lemma 8

$$S_n(a_1p^{n-1}, \dots, a_n)$$
$$= \begin{cases} p^{n(n-1)}S_1(b_1, \dots, b_n) & \text{if} \quad a_\nu = p^{\nu-1}b_\nu \ (\nu = 1, 2, \dots, n), \\ 0 & \text{otherwise,} \end{cases}$$

where

$$S_1(b_1, \dots, b_n) = \sum_{x_1,\dots,x_n}^{p} e^{2\pi i \frac{b_1(x_1+\dots+x_n)+\dots+b_n(x_1^n+\dots+x_n^n)}{p}}.$$

Therefore,

$$p^{-\frac{n(n-3)}{2}} T^*(\lambda_1, \dots, \lambda_n)$$
$$= \sum_{b_1,\dots,b_n=1}^{p} S_1(b_1, \dots, b_n) e^{-2\pi i \frac{b_1\lambda_1+\dots+b_n\lambda_n}{p}}$$
$$= \sum_{x_1,\dots,x_n}^{p} \sum_{b_1,\dots,b_n=1}^{p} e^{2\pi i \frac{b_1(x_1+\dots+x_n-\lambda_1)+\dots+b_n(x_1^n+\dots+x_n^n-\lambda_n)}{p}}.$$

Now, using Lemma 2, we obtain

$$T^*(\lambda_1, \dots, \lambda_n) = p^{\frac{n(n-1)}{2}} \sum_{x_1,\dots,x_n}^{p} \prod_{\nu=1}^{n} \delta_p(x_1^\nu + \dots + x_n^\nu - \lambda_\nu)$$
$$\leqslant p^{\frac{n(n-1)}{2}} \sum_{x_1,\dots,x_n=1}^{p} \prod_{\nu=1}^{n} \delta_p(x_1^\nu + \dots + x_n^\nu - \lambda_\nu),$$
$$= p^{\frac{n(n-1)}{2}} T(\lambda_1, \dots, \lambda_n),$$

where $T(\lambda_1,\dots,\lambda_n)$ is the number of solutions of the system of congruences

$$\left.\begin{array}{l} x_1+\dots+x_n\equiv\lambda_1 \\ \dots\dots\dots\dots \\ x_1^n+\dots+x_n^n\equiv\lambda_n \end{array}\right\} \pmod p, \qquad 1\leqslant x_\nu\leqslant p.$$

Hence, because $T(\lambda_1,\dots,\lambda_n)\leqslant n!$ by (82), the lemma assertion follows:

$$T^*(\lambda_1,\dots,\lambda_n)\leqslant p^{\frac{n(n-1)}{2}}T(\lambda_1,\dots,\lambda_n)\leqslant n!\,p^{n(n-1)/2}.$$

COROLLARY. *Let $T_n^*(mp^n)$ be the number of solutions of the system of congruences*

$$\left.\begin{array}{l} x_1+\dots-y_n\equiv 0 \pmod p \\ \dots\dots\dots\dots\dots\dots \\ x_1^n+\dots-y_n^n\equiv 0 \pmod{p^n} \end{array}\right\}, \qquad \begin{array}{l} 1\leqslant x_j,y_j\leqslant mp^n, \\ i\neq j\Rightarrow x_i\not\equiv x_j,\ y_i\not\equiv y_j \pmod p. \end{array}$$

Then

$$T_n^*(mp^n)\leqslant n!\,m^{2n}p^{2n^2-\frac{n(n+1)}{2}}. \tag{93}$$

Proof. Since each variable among $x_1,\dots,y_n$ runs through a complete residue system modulo p^n m times (under the additional conditions $i\neq j\Rightarrow x_i\not\equiv x_j, y_i\not\equiv y_j \pmod p$), then using the lemma we obtain

$$\begin{aligned} T_n^*(mp^n) &= m^{2n}T_n^*(p^n) \\ &= m^{2n}\sum_{y_1,\dots,y_n}^{p^n} T^*(y_1+\dots+y_n,\dots,y_1^n+\dots+y_n^n) \\ &\leqslant m^{2n}p^{n^2}n!\,p^{\frac{n(n-1)}{2}} = n!\,m^{2n}p^{2n^2-\frac{n(n+1)}{2}}. \end{aligned}$$

§ 7. Sums with exponential function

Let a be an integer, $m\geqslant 2$ and $q\geqslant 2$ be coprime positive integers. Sums of the form

$$S(P)=\sum_{x=1}^{P} e^{2\pi i\frac{aq^x}{m}}$$

are called *rational exponential sums containing an exponential function.* In the investigation of such sums we shall need some properties of the order of q for modulus m.

Let p be a prime, $m = pm_1$, τ and τ_1 be the orders of q for moduli m and m_1, respectively. We shall show, that if $\tau \neq \tau_1$ and $p\backslash m_1$, then the equality

$$\tau = p\tau_1 \tag{94}$$

holds.

Indeed, since $m_1\backslash m$, then from the congruence $q^\tau \equiv 1 \pmod{m}$ we get $q^\tau \equiv 1 \pmod{m_1}$ and, therefore, $\tau_1\backslash\tau$. On the other hand, from the congruence $q^{\tau_1} \equiv 1 \pmod{m_1}$ we obtain $q^{\tau_1} = 1 + u_1 m_1$, where u_1 is an integer and m_1 is a multiple of p by the assumption. But then

$$q^{p\tau_1} = (1 + u_1 m_1)^p \equiv 1 \pmod{m}$$

and $\tau\backslash p\tau_1$. Since $\tau_1\backslash\tau$, $\tau_1 \neq \tau$ and p is a prime, the equality (94) follows:

$$\tau\tau_1^{-1}\backslash p, \qquad \tau\tau_1^{-1} = p, \qquad \tau = p\tau_1.$$

Let now m be odd, $m = p_1^{\alpha_1} \dots p_s^{\alpha_s}$ be the prime factorization of m, τ and τ_1 be the orders of q for moduli m and $p_1 \dots p_s$, respectively.

We determine the quantities $\beta_1, \dots, \beta_s$ with the help of the conditions

$$q^{\tau_1} - 1 = u_0 p_1^{\beta_1} \dots p_s^{\beta_s} \qquad (u_0, p_1 \dots p_s) = 1. \tag{95}$$

For definiteness we suppose that in the prime factorization of m those primes, which satisfy the inequality $\alpha_\nu > \beta_\nu$, are put at the first r places $(0 \leqslant r \leqslant s)$, so $\alpha_\nu > \beta_\nu$ under $\nu \leqslant r$ and $\alpha_\nu \leqslant \beta_\nu$ under $\nu > r$. Further let

$$m_1 = p_1^{\beta_1} \dots p_r^{\beta_r} p_{r+1}^{\alpha_{r+1}} \dots p_s^{\alpha_s}.$$

From the definition of τ_1 and the equality (95) it follows that the order of q for modulus m_1 is equal to τ_1 and

$$q^{\tau_1} = 1 + u_1 m_1, \qquad (u_1, p_1 \dots p_r) = 1.$$

Let us show the validity of the equalities

$$q^\tau = 1 + u\,m, \qquad (u, p_1 \dots p_r) = 1 \quad \text{and} \quad \tau = \frac{m}{m_1}\tau_1. \tag{96}$$

Indeed, let $m_2 = pm_1$, where p is any number among the primes $p_1, \dots, p_r$. Let τ_2 denote the order of q for modulus m_2. Obviously $\tau_2 \neq \tau_1$ (for otherwise we would have $m_2\backslash q^{\tau_1} - 1$ and $pm_1\backslash u_1 m_1$, which contradicts the condition $(u_1, p_1 \dots p_r) = 1$). Since, besides that, $p\backslash m_1$, then by (94) $\tau_2 = p\tau_1$. But then

$$\begin{aligned} q^{p\tau_1} &= (1 + u_1 m_1)^p \equiv 1 + u_1 m_2 \pmod{p_1 \dots p_r m_2}, \\ q^{\tau_2} &= 1 + u_2 m_2, \end{aligned}$$

where $u_2 \equiv u_1 \pmod{p_1 \dots p_r}$, and, therefore, $(u_2, p_1 \dots p_r) = 1$. Thus

$$q^{\tau_2} = 1 + u_2 m_2, \qquad (u_2, p_1 \dots p_r) = 1 \quad \text{and} \quad \tau_2 = \frac{m_2}{m_1}\tau_1.$$

Repeating this process $\alpha_\nu - \beta_\nu$ times with p being equal to each p_ν $(\nu = 1, 2, \dots, r)$, we obtain the equality (96).

THEOREM 8. *Let $m \geqslant 2$, p be a prime, $m = pm_1$, τ and τ_1 be the orders of q for moduli m and m_1, respectively. If $\tau \neq \tau_1$ and $p^2 \backslash m$, then under any a not divisible by p*

$$\sum_{x=1}^{\tau} e^{2\pi i \frac{aq^x}{m}} = 0. \tag{97}$$

Proof. Let T denote the number of solutions of the congruence

$$q^x \equiv q^y \pmod m, \qquad 1 \leqslant x, y \leqslant \tau.$$

Using Lemma 2, we obtain

$$T = \sum_{x,y=1}^{\tau} \delta_m(q^x - q^y) = \frac{1}{m} \sum_{a=1}^{m} \sum_{x,y=1}^{\tau} e^{2\pi i \frac{a(q^x - q^y)}{m}}$$
$$= \frac{1}{m} \sum_{a=1}^{m} \left| \sum_{x=1}^{\tau} e^{2\pi i \frac{aq^x}{m}} \right|^2.$$

On the other hand, obviously, $T = \tau$. But then

$$\sum_{a=1}^{m} \left| \sum_{x=1}^{\tau} e^{2\pi i \frac{aq^x}{m}} \right|^2 = mT = m\tau. \tag{98}$$

Therefore, by (94)

$$\sum_{\substack{a=1 \\ (a,p)=1}}^{m} \left| \sum_{x=1}^{\tau} e^{2\pi i \frac{aq^x}{m}} \right|^2$$
$$= \sum_{a=1}^{m} \left| \sum_{x=1}^{\tau} e^{2\pi i \frac{aq^x}{m}} \right|^2 - \sum_{\substack{a=1 \\ (a,p)=p}}^{m} \left| \sum_{x=1}^{\tau} e^{2\pi i \frac{aq^x}{m}} \right|^2$$
$$= m\tau - \sum_{a_1=1}^{m_1} \left| \sum_{x=1}^{p\tau_1} e^{2\pi i \frac{a_1 q^x}{m_1}} \right|^2 = m\tau - p^2 m_1 \tau_1 = 0.$$

Hence for any a not divisible by p we obtain the theorem assertion

$$\sum_{x=1}^{\tau} e^{2\pi i \frac{aq^x}{m}} = 0.$$

Let us show that if at least one of the theorem conditions

$$\tau \neq \tau_1 \quad \text{and} \quad p^2 \backslash m$$

is not satisfied, then the sum (97) might be not equal to zero.

Indeed, let $p > 2$ be an arbitrary prime, $m = 2p$, (a, m)=1, and q be a primitive root of $2p$. Then we have

$$\sum_{x=1}^{\tau} e^{2\pi i \frac{aq^x}{m}} = \sum_{\substack{x=1 \\ x \neq \frac{p+1}{2}}}^{p} e^{2\pi i \frac{a(2x-1)}{2p}} = -e^{2\pi i \frac{a}{2}} = 1.$$

In this example the condition $\tau \neq \tau_1$ is obviously satisfied, but $p^2 \nmid 2p$, and the second condition is violated.

Let now $m = p^2$, g be a primitive root of p^2, and $q = g^p$. Then $\tau = p - 1$ and, using the equality (98), we obtain

$$\max_{(a,p)=1} \left| \sum_{x=1}^{\tau} e^{2\pi i \frac{aq^x}{m}} \right|^2 \geqslant \frac{1}{p(p-1)} \sum_{\substack{a=1 \\ (a,p)=1}}^{p^2} \left| \sum_{x=1}^{p-1} e^{2\pi i \frac{ag^{px}}{p^2}} \right|^2$$

$$= \frac{1}{p(p-1)} \sum_{a=1}^{p^2} \left| \sum_{x=1}^{p-1} e^{2\pi i \frac{ag^{px}}{p^2}} \right|^2 - \frac{1}{p(p-1)} \sum_{a_1=1}^{p} \left| \sum_{x=1}^{p-1} e^{2\pi i \frac{a_1 g^x}{p}} \right|^2 = p - 1,$$

$$\max_{(a,p)=1} \left| \sum_{x=1}^{\tau} e^{2\pi i \frac{aq^x}{m}} \right| \leqslant \sqrt{p-1}.$$

In this case the condition $p^2 \backslash m$ was fulfilled, but $\tau = \tau_1$.

Another form of conditions, under which the complete sums $S(\tau)$ vanish, is shown in the following theorem.

THEOREM 9. *Let $m = p_1^{\alpha_1} \dots p_s^{\alpha_s}$ be prime factorization of odd m, τ be the order of q for modulus m and the quantities $\beta_1, \dots, \beta_s$ be determined by the equality* (95). *If there exists ν such that $\alpha_\nu > \beta_\nu$ and $a \not\equiv 0 \pmod{p_\nu^{\alpha_\nu - \beta_\nu}}$, then*

$$\sum_{x=1}^{\tau} e^{2\pi i \frac{aq^x}{m}} = 0.$$

Proof. Chose that value ν, which satisfies the conditions

$$\alpha_\nu > \beta_\nu, \qquad a \not\equiv 0 \pmod{p_\nu^{\alpha_\nu - \beta_\nu}},$$

and write a in the form $a = p^{\alpha_\nu - \beta_\nu - \gamma} a_1$, where $\gamma \geqslant 1$ and $(a', p_\nu) = 1$. Let $m = p_\nu^{\alpha_\nu - \beta_\nu - \gamma} m'$, $m' = p_\nu m''$, τ' and τ'' be the orders of q for moduli m' and m'', respectively. Since $p_\nu^{\alpha_\nu} \backslash m$, then $p_\nu^{\beta_\nu + \gamma} \backslash m'$. But then $p_\nu^{\beta_\nu} \backslash m''$ and by (96)

$$\tau'' = \frac{m''}{m_1} \tau_1, \qquad \tau' = \frac{m'}{m_1} \tau_1 = \frac{m'}{m''} \tau'' = p_\nu \tau''.$$

From the divisibility of m' by $p_\nu^{\beta_\nu+\gamma}$ it follows also, that $p_\nu^2 \backslash m'$. Thus

$$m' = p_\nu m'', \qquad \tau' \neq \tau'', \qquad p_\nu^2 \backslash m' \quad \text{and} \quad (a', p_\nu) = 1.$$

Therefore, by Theorem 8

$$\sum_{x=1}^{\tau'} e^{2\pi i \frac{a'q^x}{m'}} = 0.$$

Since $q^\tau \equiv 1 \pmod{m}$ and $m' \backslash m$, then $q^\tau \equiv 1 \pmod{m'}$ and τ' is a divisor of τ. Now, using the property (26), we obtain the theorem assertion:

$$\sum_{x=1}^{\tau} e^{2\pi i \frac{aq^x}{m}} = \sum_{x=1}^{\tau} e^{2\pi i \frac{a'q^x}{m'}} = \frac{\tau}{\tau'} \sum_{x=1}^{\tau'} e^{2\pi i \frac{a'q^x}{m'}} = 0.$$

Note that the Theorem 9 requirements can be relaxed, namely, the condition of m being odd may be omitted. In order to prove that it suffices (see [32]) in determining the quantities $\beta_1, \ldots, \beta_s$ to use the equality

$$q^{(\mu+1)\tau_1} - 1 = u_0 p_1^{\beta_1} \ldots p_s^{\beta_s}, \qquad (u_0, p_1 \ldots p_s) = 1,$$

where $\mu = 1$, if $m \equiv 0 \pmod 2$, $\tau_1 \equiv 1 \pmod 2$, $q \equiv 3 \pmod 4$, and $\mu = 0$ otherwise, instead of the equality (95).

THEOREM 10. *Let $m \geqslant 2$ be an arbitrary integer, $(a, m) = 1$, $(q, m) = 1$, and τ be the order of q for modulus m. Then the estimate*

$$\left| \sum_{x=1}^{\tau} e^{2\pi i \left(\frac{aq^x}{m} + \frac{bx}{\tau} \right)} \right| \leqslant \sqrt{m} \tag{99}$$

holds under any integer b.

Proof. Since the fractional parts

$$\left\{ \frac{aq^x}{m} \right\} \quad \text{and} \quad \left\{ \frac{bx}{\tau} \right\}$$

have the same period τ, then by (28) the sum (99) is a complete exponential sum. But then under any integer z

$$\sum_{x=1}^{\tau} e^{2\pi i \left(\frac{aq^x}{m} + \frac{bx}{\tau} \right)} = \sum_{x=1}^{\tau} e^{2\pi i \left(\frac{aq^{x+z}}{m} + \frac{bx+bz}{\tau} \right)},$$

$$\left| \sum_{x=1}^{\tau} e^{2\pi i \left(\frac{aq^x}{m} + \frac{bx}{\tau} \right)} \right| = \left| \sum_{x=1}^{\tau} e^{2\pi i \left(\frac{aq^x q^z}{m} + \frac{bx}{\tau} \right)} \right|.$$

Therefore,

$$\tau\left|\sum_{x=1}^{\tau} e^{2\pi i\left(\frac{aq^x}{m}+\frac{bx}{\tau}\right)}\right|^2 = \sum_{z=1}^{\tau}\left|\sum_{x=1}^{\tau} e^{2\pi i\left(\frac{aq^x q^z}{m}+\frac{bx}{\tau}\right)}\right|^2 \leqslant \sum_{z=1}^{m}\left|\sum_{x=1}^{\tau} e^{2\pi i\left(\frac{aq^x z}{m}+\frac{bx}{\tau}\right)}\right|^2.$$

Hence the theorem assertion follows, because the congruence

$$q^x \equiv q^y \pmod m, \qquad 1 \leqslant x, y \leqslant \tau,$$

is satisfied for $x = y$ only:

$$\left|\sum_{x=1}^{\tau} e^{2\pi i\left(\frac{aq^x}{m}+\frac{bx}{\tau}\right)}\right|^2 \leqslant \frac{1}{\tau}\sum_{z=1}^{m}\left|\sum_{x=1}^{\tau} e^{2\pi i\left(\frac{aq^x z}{m}+\frac{bx}{\tau}\right)}\right|^2$$

$$= \frac{1}{\tau}\sum_{x,y=1}^{\tau} e^{2\pi i\frac{b(x-y)}{\tau}} \sum_{z=1}^{m} e^{2\pi i\frac{(q^x-q^y)z}{m}}$$

$$= \frac{m}{\tau}\sum_{x,y=1}^{\tau} e^{2\pi i\frac{b(x-y)}{\tau}} \delta_m(q^x - q^y) = m,$$

$$\left|\sum_{x=1}^{\tau} e^{2\pi i\left(\frac{aq^x}{m}+\frac{bx}{\tau}\right)}\right|^2 \leqslant \sqrt{m}.$$

§ 8. Distribution of digits in complete period of periodic fractions

Let $\frac{a}{m}$ be an irreducible fraction and $q \geqslant 2$ an arbitrary integer prime to m. In writing the q-adic expansion of the number $\frac{a}{m}$, the following infinite pure recurring "decimal" to the base q arises:

$$\frac{a}{m} = \left[\frac{a}{m}\right] + 0.\gamma_1\gamma_2\ldots\gamma_x\ldots, \qquad \gamma_{x+\tau} = \gamma_x \quad (x \geqslant 1), \tag{100}$$

with a period τ being equal to the order, to which q belongs for modulus m.

Let $N_m^{(P)}(\delta_1\ldots\delta_n)$ denote the number of the times that the following equation is satisfied:

$$\gamma_{x+1}\cdots\gamma_{x+n} = \delta_1\ldots\delta_n \qquad (x = 0, 1, \ldots, P-1),$$

where $P \leqslant \tau$ and $\delta_1\ldots\delta_n$ is an arbitrary fixed n-digited number in the scale of q. In other words, $N_m^{(P)}(\delta_1\ldots\delta_n)$ is the number of occurrences of the given block $\delta_1\ldots\delta_n$ of digits of length n among the first P blocks

$$\gamma_1\cdots\gamma_n,\ \gamma_2\cdots\gamma_{n+1},\ \ldots,\ \gamma_P\cdots\gamma_{P+n-1},$$

formed by successive digits of the expansion (100).

The question about the nature of the distribution of digits in the period of the fraction $\frac{a}{m}$ is closely connected with properties of rational exponential sums containing exponential function. This connection is based on the possibility to represent the quantity $N_m^{(P)}(\delta_1 \dots \delta_n)$ in terms of the number of solutions of the congruence

$$aq^x \equiv y + b \pmod{m}, \qquad 0 \leqslant x < P, \quad 1 \leqslant y \leqslant h, \tag{101}$$

where b and h depend on a choice of the block of digits $\delta_1 \dots \delta_n$. We denote the number of solutions of the congruence (101) by $T_m^{(P)}(b, h)$.

LEMMA 10. *Let quantities t, b, and h be defined by the equalities*

$$0.\delta_1 \dots \delta_n = \frac{t}{q^n}, \qquad b = \left[\frac{tm}{q^n}\right], \qquad h = \left[\frac{(t+1)m}{q^n}\right] - \left[\frac{tm}{q^n}\right].$$

Then

$$N_m^{(P)}(\delta_1 \dots \delta_n) = T_m^{(P)}(b, h).$$

Proof. Let x be any solution of the equation

$$\gamma_{x+1} \dots \gamma_{x+n} = \delta_1 \dots \delta_n \qquad (0 \leqslant x < P). \tag{102}$$

Then we obtain from (100)

$$\begin{aligned}\left\{\frac{aq^x}{m}\right\} &= 0.\gamma_{x+1} \dots \gamma_{x+n} \dots = 0.\gamma_{x+1} \dots \gamma_{x+n} + \frac{\theta_x}{q^n} \\ &= 0.\delta_1 \dots \delta_n + \frac{\theta_x}{q^n} = \frac{t + \theta_x}{q^n},\end{aligned}$$

where $0 < \theta_x < 1$. Hence it is plain that the equality (102) is satisfied for those and only those x, for which

$$\frac{t}{q^n} < \left\{\frac{aq^x}{m}\right\} < \frac{t+1}{q^n}, \qquad 0 \leqslant x < P. \tag{103}$$

Since from the definition of b and h it follows that

$$\frac{b}{m} \leqslant \frac{t}{q^n} < \frac{b+1}{m} \quad \text{and} \quad \frac{b+h}{m} \leqslant \frac{t+1}{q^n} < \frac{b+h+1}{m},$$

then the inequalities (103) are equivalent to the inequalities

$$\frac{b}{m} < \left\{\frac{aq^x}{m}\right\} \leqslant \frac{b+h}{m}, \qquad 0 \leqslant x < P. \tag{104}$$

We use y to denote the least non-negative residue of aq^x to modulus m. Then $\frac{aq^x}{m} = \frac{y}{m}$ and the inequalities (104) are satisfied for those and only those x, which satisfy the congruence

$$aq^x \equiv y \pmod m, \qquad 0 \leqslant x < P, \quad b < y \leqslant b + h,$$

or, that is just the same, the congruence

$$aq^x \equiv y + b \pmod m, \qquad 0 \leqslant x < P, \quad 1 \leqslant y \leqslant h. \tag{105}$$

But then the number of solutions of the congruence (105) coincides with the number of solutions of the equation (102). The lemma is proved.

Let m be odd, $m = p_1^{\alpha_1} \dots p_s^{\alpha_s}$ the prime factorization of m, τ_1 the order of q for modulus $p_1 \dots p_s$ and quantities $\beta_1, \dots, \beta_s$ be determined as in § 7 with the help of the conditions

$$q^{\tau_1} - 1 = u_0 p_1^{\beta_1} \dots p_s^{\beta_s}, \qquad (u_0, p_1 \dots p_s) = 1.$$

We assume that $\alpha_\nu > \beta_\nu$ under $\nu \leqslant r$ and $\alpha_\nu \leqslant \beta_\nu$ under $\nu > r$ $(0 \leqslant r \leqslant s)$. Choose

$$m_1 = p_1^{\beta_1} \dots p_r^{\beta_r} p_{r+1}^{\alpha_{r+1}} \dots p_s^{\alpha_s}.$$

Then the order of q for modulus m_1 should be equal to τ_1 and $\tau = \frac{m}{m_1}\tau_1$ by (96).

LEMMA 11. *Let $b \equiv b_1 \pmod{m_1}$, $h \equiv h_1 \pmod{m_1}$, and $h \geqslant h_1$. Then*

$$T_m^{(\tau)}(b, h) = \frac{h - h_1}{m_1}\tau_1 + T_{m_1}^{(\tau_1)}(b_1, h_1).$$

Proof. Using Lemma 2 we get

$$\begin{aligned} T_m^{(\tau)}(b, h) &= \sum_{x=0}^{\tau-1} \sum_{y=1}^{h} \delta_m(aq^x - y - b) \\ &= \frac{1}{m} \sum_{z=1}^{m} \left(\sum_{y=1}^{h} e^{-2\pi i \frac{(y+b)z}{m}} \right) \left(\sum_{x=1}^{\tau} e^{2\pi \frac{azq^x}{m}} \right). \end{aligned}$$

By Theorem 9 the inner sum of the right-hand side of this equality may not vanish only for values z, which satisfy the congruences

$$z \equiv 0 \pmod{p_\nu^{\alpha_\nu - \beta_\nu}}, \qquad \nu = 1, 2, \dots, r,$$

i.e., for

$$z = p_1^{\alpha_1 - \beta_1} \dots p_r^{\alpha_r - \beta_r} z_1 = \frac{m}{m_1} z_1.$$

Therefore, using $\tau = \frac{m}{m_1}\tau_1$ and $b \equiv b_1 \pmod{m_1}$, we obtain

$$T_m^{(\tau)}(h) = \frac{1}{m}\sum_{z_1=1}^{m_1}\left(\sum_{y=1}^{h} e^{-2\pi i\frac{(y+b)z_1}{m_1}}\right)\left(\sum_{x=1}^{\tau} e^{2\pi i\frac{az_1q^x}{m_1}}\right)$$

$$= \frac{1}{m_1}\sum_{z_1=1}^{m_1}\left(\sum_{y=1}^{h} e^{-2\pi i\frac{(y+b)z_1}{m_1}}\right)\left(\sum_{x=1}^{\tau_1} e^{2\pi i\frac{az_1q^x}{m_1}}\right)$$

$$= \sum_{x=1}^{\tau_1}\sum_{y=1}^{h}\delta_{m_1}(aq^x - y - b) = \sum_{x=1}^{\tau_1}\sum_{y=1}^{h}\delta_{m_1}(aq^x - y - b_1).$$

Since the difference $h - h_1$ is a multiple of m_1 and, therefore,

$$\sum_{y=h_1+1}^{h}\delta_{m_1}(aq^x - y - b_1) = \sum_{y=1}^{h-h_1}\delta_{m_1}(y) = \frac{h-h_1}{m_1},$$

we obtain the Lemma assertion:

$$T_m^{(\tau)}(h) = \sum_{x=1}^{\tau_1}\sum_{y=1}^{h_1}\delta_{m_1}(aq^x - y - b_1) + \sum_{x=1}^{\tau_1}\sum_{y=h_1+1}^{h}\delta_{m_1}(aq^x - y - b_1)$$

$$= \frac{h-h_1}{m_1}\tau_1 + T_{m_1}^{(\tau_1)}(b_1, h_1).$$

Let us consider the question concerning the distribution of blocks of digits in the complete period of the fraction $\frac{a}{m}$. Since there exist q^n distinct blocks $\delta_1 \ldots \delta_n$, then the mean value of the number of occurrences of a given block of n digits equals $\frac{1}{q^n}\tau$.

THEOREM 11. *Let R_n be the deviation of the quantity $N_m^{(\tau)}(\delta_1 \ldots \delta_n)$ from its mean value:*

$$N_m^{(\tau)}(\delta_1 \ldots \delta_n) = \frac{1}{q^n}\tau + R_n.$$

Then under odd m, any $n \geqslant 1$, and any choice of the digits $\delta_1 \ldots \delta_n$ the following estimate is valid:

$$|R_n| \leqslant \left(1 - \frac{1}{q^n}\right)\tau_1, \tag{106}$$

where τ_1 is the order of q for modulus being equal to the product of primes entering into the prime factorization of m.

Proof. By Lemma 10

$$N_m^{(\tau)}(\delta_1 \ldots \delta_n) = T_m^{(\tau)}(b, h),$$

where $T_m^{(\tau)}(b, h)$ is the number of solutions of the congruence

$$aq^x \equiv y + b \pmod{m}, \qquad 0 \leqslant x < \tau, \quad 1 \leqslant y \leqslant h,$$

and the quantities b and h are determined with the help of the equalities

$$0.\delta_1 \ldots \delta_n = \frac{t}{q^n}, \qquad b = \left[\frac{tm}{q^n}\right], \qquad h = \left[\frac{(t+1)m}{q^n}\right] - \left[\frac{tm}{q^n}\right].$$

Let h_1 denote the least non-negative residue of h to modulus m_1:

$$h \equiv h_1 \pmod{m_1}, \qquad 0 \leqslant h < m_1.$$

Observing that $\frac{h-h_1}{m_1}$ is an integer and

$$h = \frac{m}{q^n} - \left\{\frac{(t+1)m}{q^n}\right\} + \left\{\frac{tm}{q^n}\right\},$$

we obtain

$$\begin{aligned}\frac{h-h_1}{m_1} &= \left[\frac{m}{m_1 q^n}\right] - \left[\frac{m}{m_1 q^n} - \frac{h-h_1}{m_1}\right] \\ &= \frac{m}{m_1 q^n} - \left\{\frac{m}{m_1 q^n}\right\} \\ &\quad - \left[\frac{h_1}{m_1} + \frac{1}{m_1}\left(\left\{\frac{(t+1)m}{q^n}\right\} - \left\{\frac{tm}{q^n}\right\}\right)\right].\end{aligned}$$

But then it follows from Lemma 11 by virtue of the equality $\tau = \frac{m}{m_1}\tau_1$, that

$$\begin{aligned}T_m^{(\tau)}(b,h) &= \frac{h-h_1}{m_1}\tau_1 + T_{m_1}^{(\tau_1)}(b,h_1) \\ &= \frac{\tau}{q^n} - \left\{\frac{m}{m_1 q^n}\right\}\tau_1 + T_{m_1}^{(\tau_1)}(b,h_1) \\ &\quad - \left[\frac{h_1}{m_1} + \frac{1}{m_1}\left(\left\{\frac{(t+1)m}{q^n}\right\} - \left\{\frac{tm}{q^n}\right\}\right)\right]\tau_1.\end{aligned}$$

Since $0 \leqslant h_1 \leqslant m_1 - 1$, then

$$\begin{aligned}&\left[\frac{h_1}{m_1} + \frac{1}{m_1}\left(\left\{\frac{(t+1)m}{q^n}\right\} - \left\{\frac{tm}{q^n}\right\}\right)\right] \\ &\qquad = \begin{cases} -1 & \text{if} \quad h_1 = 0 \quad \text{and} \quad \left\{\frac{(t+1)m}{q^n}\right\} < \left\{\frac{tm}{q^n}\right\}, \\ 0 & \text{otherwise}\end{cases}\end{aligned}$$

and therefore

$$\begin{aligned}T_m^{(\tau)}(b,h) - \frac{1}{q^n}\tau &= T_{m_1}^{(\tau_1)}(b_1,h_1) - \left\{\frac{m}{m_1 q^n}\right\}\tau_1 \\ &+ \begin{cases} \tau_1 & \text{if} \quad h_1 = 0 \quad \text{and} \quad \left\{\frac{(t+1)m}{q^n}\right\} < \left\{\frac{tm}{q^n}\right\}, \\ 0 & \text{otherwise.}\end{cases}\end{aligned}$$

Since

$$0 \leqslant T_{m_1}^{(\tau_1)}(b, h_1) \leqslant \tau_1 \quad \text{and} \quad T_{m_1}^{(\tau_1)}(b, 0) = 0,$$

we obtain the theorem assertion:

$$|R_n| = \left| N_m^{(\tau)}(\delta_1 \dots \delta_n) - \frac{1}{q^n}\tau \right|$$
$$= \left| T_m^{(\tau)}(b, h) - \frac{1}{q^n}\tau \right| \leqslant \left(1 - \frac{1}{q^n}\right)\tau_1.$$

It is easy to ascertain that the estimate (106) can not be substantially improved. Indeed, let $q = 2$, $\tau > 1$, $a = 1$, and $m = 2^\tau - 1$. The 2-adic expansion of $\frac{a}{m}$ has period τ

$$\frac{a}{m} = \frac{1}{2^\tau - 1} = 0.(0 \dots 01)0 \dots 01 \dots .$$

Choose $\delta_1 \dots \delta_n = 0 \dots 0$. Then we get $N_m^{(\tau)}(\delta_1 \dots \delta_n) = \tau - n$ and

$$R_n = \tau - n - \frac{1}{q^n}\tau \geqslant \left(1 - \frac{1}{q^n}\right)\tau_1 - n.$$

Let us note also, that under $\tau_1 = 1$ it follows from Theorem 11, that

$$N_m^{(\tau)}(\delta_1 \dots \delta_n) = \frac{1}{q^n}\tau + \theta_n,$$

where $|\theta_n| < 1$. It is so, for example, under $q = 4$ and $m = 3^\alpha$. In general, for $m = {p_1}^{\alpha_1} \dots {p_s}^{\alpha_s}$ under fixed primes p_ν and growing α_ν, the magnitude τ_1 is bounded and the following asymptotic formula is valid by (106):

$$N_m^{(\tau)}(\delta_1 \dots \delta_n) = \frac{1}{q^n}\tau + O(1).$$

Now, let us establish the correlation between the occurrence of a given block of digits in period of the fractions

$$\frac{a}{m} = \left[\frac{a}{m}\right] + 0.\gamma_1\gamma_2 \dots \gamma_x \dots \qquad (\gamma_{x+\tau} = \gamma_x)$$

and

$$\frac{a}{m_1} = \left[\frac{a}{m_1}\right] + 0.\gamma_1'\gamma_2' \dots \gamma_x' \dots \qquad ({\gamma'}_{x+\tau_1} = {\gamma'}_x),$$

where the quantity m_1 is determined as in Lemma 11.

THEOREM 12. *If $q^{n_0}\backslash m - m_1$ under a certain $n_0 \geqslant 1$, then*

$$N_m^{(\tau)}(\delta_1 \dots \delta_n) = \frac{\tau - \tau_1}{q^n} + N_{m_1}^{(\tau_1)}(\delta_1 \dots \delta_n)$$

under any $n \leqslant n_0$ and any choice of a block of digits $\delta_1 \dots \delta_n$.

Proof. Determine integers t, b, h, b_1, and h_1 with the help of the equalities

$$0.\delta_1 \dots \delta_n = \frac{t}{q^n}, \qquad b = \left[\frac{tm}{q^n}\right], \qquad h = \left[\frac{(t+1)m}{q^n}\right] - \left[\frac{tm}{q^n}\right],$$
$$b_1 = \left[\frac{tm_1}{q^n}\right], \qquad h_1 = \left[\frac{(t+1)m_1}{q^n}\right] - \left[\frac{tm_1}{q^n}\right].$$

Then obviously

$$q^n(b - b_1) = t(m - m_1) - q^n\left(\left\{\frac{tm}{q^n}\right\} - \left\{\frac{tm_1}{q^n}\right\}\right),$$
$$q^n(h - h_1) = m - m_1$$
$$- q^n\left(\left\{\frac{(t+1)m}{q^n}\right\} - \left\{\frac{(t+1)m_1}{q^n}\right\} + \left\{\frac{tm_1}{q^n}\right\} - \left\{\frac{tm}{q^n}\right\}\right).$$

Using the congruence $m \equiv m_1 \pmod{q^n}$, which is satisfied under $n \leqslant n_0$, we get

$$\left\{\frac{tm}{q^n}\right\} = \left\{\frac{tm_1}{q^n}\right\}, \qquad \left\{\frac{(t+1)m}{q^n}\right\} = \left\{\frac{(t+1)m_1}{q^n}\right\}.$$

Therefore,

$$q^n\,(b - b_1) = t\,(m - m_1), \qquad q^n\,(h - h_1) = m - m_1, \tag{107}$$

and, since $m_1 \backslash m$ and $(q, m_1) = 1$,

$$b \equiv b_1 \pmod{m_1}, \qquad h \equiv h_1 \pmod{m_1}.$$

But then, according to Lemma 11,

$$T_m^{(\tau)}(b, h) = \frac{h - h_1}{m_1}\tau_1 + T_{m_1}^{(\tau_1)}(b_1, h_1).$$

Multiply the second equality of (107) by $\frac{\tau_1}{m_1 q^n}$. Then, observing that $\tau = \frac{m}{m_1}\tau_1$, we obtain

$$\frac{h - h_1}{m_1}\tau_1 = \frac{m - m_1}{m_1 q^n}\tau_1 = \frac{\tau - \tau_1}{q^n},$$

and, therefore,

$$T_m^{(\tau)}(b, h) = \frac{\tau - \tau_1}{q^n} + T_{m_1}^{(\tau_1)}(b_1, h_1).$$

Hence, applying Lemma 10, we get the theorem assertion

$$N_m^{(\tau)}(\delta_1 \dots \delta_n) = \frac{\tau - \tau_1}{q^n} + N_{m_1}^{(\tau_1)}(\delta_1 \dots \delta_n).$$

Let us notice particularly the case $q = 2$, $m = p^\alpha$, and $m_1 = p^\beta$, where p is a prime greater than 2. Suppose further $\beta = 1$ (it is so, for instance, under $p = 3, 5, 7$) and under $n \leqslant n_0$ compare the numbers of occurrences of any n-digited block in the period of 2-adic expansion of the fractions $\frac{1}{p^\alpha}$ and $\frac{1}{p}$. The former exceeds the latter by one and the same quantity (being equal to $\frac{\tau-\tau_1}{2^n}$).

So, for example, under $m = 27$ we get $m_1 = 3$, $\tau = 18$, $\tau_1 = 2$ and $n_0 = 3$. Therefore, the number of occurrences of any block of digits of length 1, 2, or 3 in the period of the fraction

$$\frac{1}{27} = 0.(000010010111101101)00\dots$$

exceeds by 8, 4, and 2, respectively, the number of occurrences of the same block of digits in the period of the fraction

$$\frac{1}{3} = 0.(01)01\dots.$$

Analogously under $m = 25$ we obtain $n_0 = 2$, and the number of occurrences of any block of digits of length 1 or 2 in the period of the fraction

$$\frac{1}{25} = 0.(00001010001111010111)00\dots$$

is 8 or 4, respectively, more than the number of occurrences of the same block of digits in the period of the fraction

$$\frac{1}{5} = 0.(0011)00\dots.$$

These relations can be observed under $p = 3, 5, 7$, $n = 1, 2, 3$, and $p^\alpha \leqslant 125$ in the table given below.

Table of values of $N_{p^\alpha}^{(\tau)}(\delta_1 \ldots \delta_n)$

$\delta_1 \ldots \delta_n$ \ $\frac{1}{p^\alpha}$	$\frac{1}{3}$	$\frac{1}{9}$	$\frac{1}{27}$	$\frac{1}{81}$	$\frac{1}{5}$	$\frac{1}{25}$	$\frac{1}{125}$	$\frac{1}{7}$	$\frac{1}{49}$
0	1	3	9	27	2	10	50	2	11
1	1	3	9	27	2	10	50	1	10
00	0	2	4	14	1	5	25	1	6
01	1	1	5	13	1	5	25	1	6
10	1	1	5	13	1	5	25	1	5
11	0	2	4	14	1	5	25	0	5
000	0	1	2	7	0	3	12	0	3
001	0	1	2	7	1	2	13	1	3
010	1	0	3	6	0	3	12	1	3
011	0	1	2	7	1	2	13	0	2
100	0	1	2	7	1	2	13	0	3
101	1	0	3	6	0	3	12	1	2
110	0	1	2	7	1	2	13	0	2
111	0	1	2	7	0	3	12	0	3

§ 9. Exponential sums with recurrent function

Let us consider functions $\psi(x)$ satisfying the linear difference equation with constant coefficients

$$\psi(x) = a_1\psi(x-1) + \ldots + a_n\psi(x-n) \qquad (x > n). \tag{108}$$

It is known (see, for example, [11]) that any function $\psi(x)$ determined by the recurrence equality (108) can be represented in the form

$$\psi(x) = \mathcal{P}_1(x)\lambda_1^x + \ldots + \mathcal{P}_r(x)\lambda_r^x,$$

where $r \leqslant n$, $\lambda_1, \ldots, \lambda_r$ are distinct roots of the characteristic equation

$$\lambda^n = a_1\lambda^{n-1} + \ldots + a_n, \tag{109}$$

and $\mathcal{P}_1(x), \ldots, \mathcal{P}_r(x)$ are polynomials whose degrees are unity less than the multiplicity of the corresponding roots of the equation (109). In particular, if the characteristic equation has no multiple roots, then

$$\psi(x) = C_1\lambda_1^x + \ldots + C_n\lambda_n^x, \tag{110}$$

where $C_1, \ldots, C_n$ are constants depending upon the choice of initial values of the function $\psi(x)$. If coefficients of the equation (108) and initial values $\psi(1), \ldots, \psi(n)$

are integers, then, obviously, under any positive integer x the function $\psi(x)$ takes on integer values.

Let $m > 1$, $(a_n, m) = 1$, and at least one of the initial values $\psi(1), \ldots, \psi(n)$ be not a multiple of m.

In the equation (108) we replace x by $x + n$ and transit to the congruence to the modulus m:

$$\psi(x+n) \equiv a_1\psi(x+n-1) + \ldots + a_n\psi(x) \pmod{m}. \tag{111}$$

Since $(a_n, m) = 1$, so in this congruence $\psi(x)$ can be expressed in terms of $\psi(x+1), \ldots, \psi(x+n)$ and, setting $x = 0, -1, -2, \ldots$, we may extend the function $\psi(x)$ for integers $x \leqslant 0$.

A function $\psi(x)$ determined for integers x by the congruence (111) and initial values $\psi(1), \ldots, \psi(n)$ (see [21]) is called *a recurrent function of the n-th order to the modulus m*, and the sum

$$S(P) = \sum_{x=1}^{P} e^{2\pi i \frac{\psi(x)}{m}}$$

an exponential sum with a recurrent function. It is easily seen that under $n = 1$ these sums coincide with considered in § 7 sums with an exponential function.

Let us show that a sequence of least non-negative residues of the function $\psi(x)$ to modulus m is periodic and that its least period does not exceed the quantity $m^n - 1$.

In fact, let us denote the least non-negative residue of $\psi(x)$ to modulus m by γ_x:

$$\psi(x) \equiv \gamma_x \pmod{m}, \qquad 0 \leqslant \gamma_x \leqslant m-1.$$

Then by virtue of (111)

$$\gamma_{x+n} \equiv a_1\gamma_{x+n-1} + \ldots + a_n\gamma_x \pmod{m}. \tag{112}$$

Consider blocks of n digits with respect to the base m

$$\gamma_{x+1} \cdots \gamma_{x+n} \qquad (x = 0, 1, \ldots, m^n). \tag{113}$$

Since the number of distinct blocks of n digits is equal to m^n, then among the blocks (113) there exist two identical blocks

$$\gamma_{x_1+1} \cdots \gamma_{x_1+n} = \gamma_{x_2+1} \cdots \gamma_{x_2+n} \qquad (x_2 > x_1). \tag{114}$$

We determine τ by the equality $\tau = x_2 - x_1$ and will show that under any $x \geqslant x_1$

$$\gamma_{x+1} \cdots \gamma_{x+n} = \gamma_{x+\tau+1} \cdots \gamma_{x+\tau+n}. \tag{115}$$

In fact, under $x = x_1$ this equality is fulfilled by virtue of (114). Apply the induction. Let us suppose that the equality (115) holds for a certain $x \geqslant x_1$. In the congruence (112) we replace x by $x + \tau + 1$. Then using the induction hypothesis we obtain

$$\begin{aligned}\gamma_{x+\tau+n+1} &\equiv a_1\gamma_{x+\tau+n} + \ldots + a_n\gamma_{x+\tau+1} \\ &= a_1\gamma_{x+n} + \ldots + a_n\gamma_{x+1} \equiv \gamma_{x+n+1} \pmod m,\end{aligned}$$

and, therefore, $\gamma_{x+\tau+n+1} = \gamma_{x+n+1}$. But then

$$\gamma_{x+2} \ldots \gamma_{x+n+1} = \gamma_{x+\tau+2} \ldots \gamma_{x+\tau+n+1},$$

hence the equality (115) is proved for any $x \geqslant x_1$. By means of such considerations we get this equality for $x < x_1$ as well (but in this case, γ_x should be expressed from the congruence (112) in terms of $\gamma_{x+1}, \ldots, \gamma_{x+n}$ beforehand, and that could be done because of $(a_n, m) = 1$).

Hence it follows that the least non-negative residues of the function $\psi(x)$ have a period τ, where $1 \leqslant \tau \leqslant m^n$. Let us assume that the least period is equal to m^n. Then any block of n digits should occur among the blocks (113), and, in particular, the block $0 \ldots 0$ being formed by zeros only is present among them. But then by (112) all terms of the sequence of the residues will equal zero and its least period equal 1, that contradicts to the assumption. Therefore, the least period of the function $\psi(x)$ does not exceed $m^n - 1$.

Henceforward we let τ denote the least period of the sequence of the least non-negative residues of the function $\psi(x)$ to modulus m. It is easily seen that τ is the period of fractional parts of the function $\frac{\psi(x)}{m}$:

$$\left\{\frac{\psi(x+\tau)}{m}\right\} = \frac{\gamma_{x+\tau}}{m} = \frac{\gamma_x}{m} = \left\{\frac{\psi(x)}{m}\right\}.$$

Therefore, the sum

$$S(\tau) = \sum_{x=1}^{\tau} e^{2\pi i \frac{\psi(x)}{m}}$$

is a complete exponential sum. Since under integer a

$$\left\{\frac{a(x+\tau)}{\tau}\right\} = \left\{\frac{ax}{\tau}\right\},$$

then by (28) under any integer a the sum

$$\sum_{x=1}^{\tau} e^{2\pi i \left(\frac{\psi(x)}{m} + \frac{ax}{\tau}\right)}$$

is a complete sum as well.

Let $\psi_1(x), \ldots, \psi_n(x)$ be recurrent functions satisfying the equation (108) and determined by initial values

$$\psi_j(x) = \begin{cases} 1 & \text{if} \quad x = j, \\ 0 & \text{if} \quad 1 \leqslant x \leqslant n, \quad x \neq j \quad (j = 1, 2, \ldots, n). \end{cases}$$

It is easy to show that

$$\psi(x+z) = \psi(z+1)\psi_1(x) + \ldots + \psi(z+n)\psi_n(x). \tag{116}$$

In fact by virtue of the linearity of the equation (108), any linear combination of its solutions is a solution too. In particular, the sum in the right-hand side of the equality (116) is a solution of the equation (108). From the definition of the functions $\psi_j(x)$ it is seen that under $x = 1, 2, \ldots, n$ the initial values of this sum are equal to $\psi(z+1), \psi(z+2), \ldots, \psi(z+n)$, respectively. The solution $\psi(x+z)$ has the same initial values. But solutions, which have the same initial values, coincide. Hence the equality (116) is proved.

THEOREM 13. *Let $\psi(x)$ be a recurrent function of the n-th order to the modulus m, τ be its least period, and $P \leqslant \tau$. Then we have the estimates*

$$\left| \sum_{x=1}^{\tau} e^{2\pi i \frac{\psi(x)}{m}} \right| \leqslant m^{\frac{n}{2}}, \qquad \left| \sum_{x=1}^{P} e^{2\pi i \frac{\psi(x)}{m}} \right| \leqslant m^{\frac{n}{2}}(1 + n \log m). \tag{117}$$

Proof. Since under an integer a the sum

$$S_a(\tau) = \sum_{x=1}^{\tau} e^{2\pi i \left(\frac{\psi(x)}{m} + \frac{ax}{\tau} \right)}$$

is a complete sum, then under any integer z

$$S_a(\tau) = \sum_{x=1}^{\tau} e^{2\pi i \left(\frac{\psi(x+z)}{m} + a\frac{x+z}{\tau} \right)},$$

$$|S_a(\tau)| = \left| \sum_{x=1}^{\tau} e^{2\pi i \left(\frac{\psi(x+z)}{m} + \frac{ax}{\tau} \right)} \right|.$$

Squaring and summing over z yields

$$\tau |S_a(\tau)|^2 = \sum_{z=0}^{\tau-1} \left| \sum_{x=1}^{\tau} e^{2\pi i \left(\frac{\psi(x+z)}{m} + \frac{ax}{\tau} \right)} \right|^2.$$

We let γ_z denote the least non-negative residue of the function $\psi(z)$ to modulus m. Then by (116)

$$\begin{aligned}\psi(x+z) &= \psi(z+1)\psi_1(x) + \ldots + \psi(z+n)\psi_n(x)\\ &\equiv \gamma_{z+1}\psi_1(x) + \ldots + \gamma_{z+n}\psi_n(x) \pmod{m},\end{aligned}$$

and, therefore,

$$\left|\sum_{x=1}^{\tau} e^{2\pi i \frac{\psi(x)}{m}}\right| = |S_a(\tau)| \leqslant m^{\frac{n}{2}}. \tag{118}$$

Since τ is the least period of γ_z to modulus m, then under $z = 0, 1, \ldots, \tau - 1$ all blocks $\gamma_{z+1} \ldots \gamma_{z+n}$ of n digits are distinct. Therefore, extending the summation to all possible blocks $z_1 \ldots z_n$ of n digits, we obtain

$$\begin{aligned}\tau|S_a(\tau)|^2 &\leqslant \sum_{z_1,\ldots,z_n=0}^{p-1}\left|\sum_{x=1}^{\tau} e^{2\pi i\left(\frac{z_1\psi_1(x)+\ldots+z_n\psi_n(x)}{m}+\frac{ax}{\tau}\right)}\right|^2\\ &= \sum_{x,y=1}^{\tau} e^{2\pi i\frac{a(x-y)}{\tau}} \sum_{z_1,\ldots,z_n=0}^{m-1} e^{2\pi i\left(\frac{\psi_1(x)-\psi_1(y)}{m}z_1+\ldots+\frac{\psi_n(x)-\psi_n(y)}{m}z_n\right)}\\ &\leqslant m^n \sum_{x,y=1}^{\tau} \delta_m\left[\psi_1(x)-\psi_1(y)\right]\ldots\delta_m\left[\psi_n(x)-\psi_n(y)\right] = m^n T,\end{aligned} \tag{119}$$

where T is the number of solutions of the system of congruences

$$\left.\begin{array}{l}\psi_1(x) \equiv \psi_1(y)\\ \ldots\ldots\ldots\ldots\\ \psi_n(x) \equiv \psi_n(y)\end{array}\right\} \pmod{m}, \qquad 1 \leqslant x, y \leqslant \tau. \tag{120}$$

Let us assume that this system has a solution with $y \neq x$. Without loss of generality, we may assume that $y > x$. Using the equality (116), we get

$$\begin{aligned}\psi(z) &= \psi(z-x+1)\psi_1(x) + \ldots + \psi(z-x+n)\psi_n(x),\\ \psi(z+y-x) &= \psi(z-x+1)\psi_1(y) + \ldots + \psi(z-x+n)\psi_n(y).\end{aligned}$$

Hence by (120) it follows that under any integer z

$$\psi(z+y-x) \equiv \psi(z) \pmod{m}.$$

But then $y-x$ is a period of γ_z, and since τ is the least period, then $y-x \geqslant \tau$, which leads to a contradiction. Thus, the congruence system (120) has no other solutions except for solutions with $y = x$ and, therefore, $T = \tau$. Now from (119), we get

$$\tau|S_a(\tau)|^2 \leqslant m^n T = m^n \tau, \qquad |S_a(\tau)| \leqslant m^{\frac{n}{2}}. \tag{121}$$

Hence under $a = 0$ the first assertion of the theorem follows:

$$\left|\sum_{x=1}^{\tau} e^{2\pi i \frac{\psi(x)}{m}}\right| = |S_0(\tau)| \leqslant m^{\frac{n}{2}}.$$

The second assertion of the theorem follows immediately from Theorem 2 and the estimate (121):

$$\left|\sum_{x=1}^{P} e^{2\pi i \frac{\psi(x)}{m}}\right| \leqslant \max_{1\leqslant a\leqslant \tau} \left|\sum_{x=1}^{\tau} e^{2\pi i \left(\frac{\psi(x)}{m}+\frac{ax}{\tau}\right)}\right| (1+\log\tau)$$
$$= \max_{1\leqslant a\leqslant \tau} |S_a(\tau)|(1+\log\tau) < m^{\frac{n}{2}}(1+n\log m).$$

Note that in the general case the order of the estimation

$$\left|\sum_{x=1}^{\tau} e^{2\pi i \frac{\psi(x)}{m}}\right| \leqslant m^{\frac{n}{2}}$$

can not be improved further. Indeed, using considerations from the theory of finite fields (see, for instance, [33]), it can be shown that under any prime $p > 2$ and positive integer $n < p$ there exist recurrent functions $\psi(x)$ of the n-th order to the modulus p with the period $\tau = p^n - 1$. Besides, roots of the corresponding characteristic equation (109) are distinct and by (110)

$$\psi(x) = C_1\lambda_1^x + \ldots + C_n\lambda_n^x. \tag{122}$$

By virtue of properties of symmetric functions there exists an equation with integral coefficients

$$\mu^n = b_1\mu^{n-1} + \ldots + b_n,$$

whose roots $\mu_1, \ldots, \mu_n$ equal $\lambda_1^2, \ldots, \lambda_n^2$, respectively, and the free term is relatively prime to p. Consider the functions $\psi(2x)$ and $\psi(2x+1)$. It follows from (122) that

$$\psi(2x) = C_1\mu_1^x + \ldots + C_n\mu_n^x,$$
$$\psi(2x+1) = C_1\lambda_1\mu_1^x + \ldots + C_n\lambda_n\mu_n^x.$$

Thus, $\psi(2x)$ and $\psi(2x+1)$ are recurrent functions of the n-th order to the modulus p satisfying the equation

$$\psi^*(x) = b_1\psi^*(x-1) + \ldots + b_n\psi^*(x-n).$$

Denote by γ_x the least non-negative residue of $\psi(x)$ to modulus p. Under $\tau_1 = \frac{1}{2}\tau$ we get* $\gamma_{2(x+\tau_1)} = \gamma_{2x+\tau} = \gamma_{2x}$ and, therefore, γ_{2x} has a period being equal to τ_1.

*Since p is odd, then $\tau_1 = \frac{p^n-1}{2}$ is an integer.

Let us assume that τ_1 is not the least period. Then we can find a positive integer $\tau_2 < \tau_1$, such that under any integer x

$$\psi(2x + 2\tau_2) \equiv \psi(2x) \pmod p. \tag{123}$$

Applying the equality (116) we obtain

$$\begin{aligned} \psi(2x + 2\tau_2) &= \psi(2x)\psi_1(2\tau_2 + 1) + \ldots + \psi(2x + n - 1)\psi_n(2\tau_2 + 1), \\ \psi(2x) &= \psi(2x)\psi_1(1) + \ldots + \psi(2x + n - 1)\psi_n(1). \end{aligned}$$

But then by (123) the congruence

$$[\psi_1(2\tau_2 + 1) - \psi_1(1)]\gamma_{2x} + \ldots + [\psi_n(2\tau_2 + 1) - \psi_n(1)]\gamma_{2x+n-1} \equiv 0 \pmod p \tag{124}$$

should be fulfilled under any integer x. From properties of solutions of the system (120), it follows that at least one of square brackets in (124) is not congruent to zero to modulus p and, therefore, the number of solutions of the congruence (124) does not exceed $p^{n-1} - 1$. On the other hand, according to the definition of the function $\psi(x)$ under $x = 1, 2, \ldots, \tau_1$ n-tuples $\gamma_{2x}, \gamma_{2x+1}, \ldots, \gamma_{2x+n-1}$ yield distinct solutions of this congruence. Since obviously

$$\tau_1 = \frac{p^n - 1}{2} > p^{n-1} - 1,$$

then we arrive at a contradiction and, therefore, τ_1 is the last period of γ_{2x}. Analogously we get that the least period of γ_{2x+1} is equal to τ_1 as well. Now, in the same fashion as in the deduction of (118), we arrive at the equalities

$$\begin{aligned} \tau_1 \left| \sum_{x=1}^{\tau_1} e^{2\pi i \frac{\psi(2x)}{p}} \right|^2 &= \sum_{z=0}^{\tau_1 - 1} \left| \sum_{x=1}^{\tau_1} e^{2\pi i \frac{\nu_{2z+1}\psi_1(2x) + \ldots + \nu_{2z+n}\psi_n(2x)}{p}} \right|^2, \\ \tau_1 \left| \sum_{x=1}^{\tau_1} e^{2\pi i \frac{\psi(2x+1)}{p}} \right|^2 &= \sum_{z=0}^{\tau_1 - 1} \left| \sum_{x=1}^{\tau_1} e^{2\pi i \frac{\nu_{2z+2}\psi_1(2x) + \ldots + \nu_{2z+1+n}\psi_n(2x)}{p}} \right|^2. \end{aligned}$$

Hence by virtue of the choice of the function $\psi(x)$, it follows that

$$\begin{aligned} &\tau_1 \left(\left| \sum_{x=1}^{\tau_1} e^{2\pi i \frac{\psi(2x)}{p}} \right|^2 + \left| \sum_{x=1}^{\tau_1} e^{2\pi i \frac{\psi(2x+1)}{p}} \right|^2 \right) \\ &= \sideset{}{'}\sum_{z_1, \ldots, z_n = 0}^{p-1} \left| \sum_{x=1}^{\tau_1} e^{2\pi i \frac{z_1\psi_1(2x) + \ldots + z_n\psi_n(2x)}{p}} \right|^2, \end{aligned} \tag{125}$$

where the sign $'$ in the sum $\sum'_{z_1,\ldots,z_n}$ indicates the deletion of n-tuple $z_1,\ldots,z_n$, formed by zeros entirely, from the range of summation. Let T_1 denote the number of solutions of the system of congruences

$$\left.\begin{array}{c}\psi_1(2x)\equiv\psi_1(2y)\\ \cdots\cdots\cdots\cdots\\ \psi_n(2x)\equiv\psi_n(2y)\end{array}\right\}\pmod p\qquad 1\leqslant x,y\leqslant\tau.$$

In the same way as in the system (120), we have $T_1=\tau_1$ and, therefore,

$$\sum_{z_1,\ldots,z_n=0}^{p-1}\left|\sum_{x=1}^{\tau_1}e^{2\pi i\frac{z_1\psi_1(2x)+\ldots+z_n\psi_n(2x)}{p}}\right|^2=p^nT_1=p^n\tau_1.$$

But then (125) can be rewritten in the form

$$\tau_1\left(\left|\sum_{x=1}^{\tau_1}e^{2\pi i\frac{\psi(2x)}{p}}\right|^2+\left|\sum_{x=1}^{\tau_1}e^{2\pi i\frac{\psi(2x+1)}{p}}\right|^2\right)=p^n\tau_1-\tau_1^2,$$

$$\left|\sum_{x=1}^{\tau_1}e^{2\pi i\frac{\psi(2x)}{p}}\right|^2+\left|\sum_{x=1}^{\tau_1}e^{2\pi i\frac{\psi(2x+1)}{p}}\right|^2=p^n-\tau_1=\frac{p^n+1}{2}. \qquad (126)$$

We determine $|S^*(\tau_1)|$ with the help of the equality

$$|S^*(\tau_1)|=\max\left(\left|\sum_{x=1}^{\tau_1}e^{2\pi i\frac{\psi(2x)}{p}}\right|,\left|\sum_{x=1}^{\tau_1}e^{2\pi i\frac{\psi(2x+1)}{p}}\right|\right).$$

Then from (126) we get

$$|S^*(\tau_1)|^2\geqslant\frac{p^n+1}{4},\qquad |S^*(\tau_1)|>\frac{1}{2}p^{\frac{n}{2}}.$$

Hence by (117) it follows that under any prime $p>2$ and any $n>1$ there exists a recurrent function of the n-th order to the modulus p such that for the exponential sum $S^*(\tau_1)$ the following estimates

$$\frac{1}{2}p^{\frac{n}{2}}<|S^*(\tau_1)|\leqslant P^{\frac{n}{2}}$$

hold.

§ 10. Sums of Legendre's symbols

Let $p > 2$ be a prime, $f(x) = a_0 + a_1x + \ldots + a_nx^n$ be a polynomial with integral coefficients, $n < p$, and $(a_n, p) = 1$. Let σ_n denote the sum of Legendre's symbols

$$\sigma_n = \sum_{x=1}^{p} \left(\frac{f(x)}{p}\right), \tag{127}$$

and T_n denote the number of solutions of the congruence

$$y^2 \equiv f(x) \pmod p. \tag{128}$$

The quantities T_n and σ_n are connected by a simple relationship

$$T_n = \sum_{x=1}^{p} \left[1 + \left(\frac{f(x)}{p}\right)\right] = p + \sigma_n. \tag{129}$$

This relationship reduces the question on the number of solutions of the congruence (128) to studies of sums of the Legendre symbols.

The sums (127) are easily evaluated for polynomials of the first and the second degree. Indeed, since Legendre's symbol $\left(\frac{x}{p}\right)$ is a periodic function with a period p and under $(a_1, p) = 1$ the linear function $a_0 + a_1x$ runs through a complete residue system modulo p, when x runs through a complete residue system modulo p, then

$$\sum_{x=1}^{p} \left(\frac{a_0 + a_1x}{p}\right) = \sum_{x=1}^{p} \left(\frac{x}{p}\right) = 0.$$

In order to evaluate the sum

$$\sigma_2 = \sum_{x=1}^{p} \left(\frac{a_0 + a_1x + a_2x^2}{p}\right),$$

we consider the congruence

$$y^2 \equiv x^2 + a \pmod p$$

and denote the number of its solutions by $T(a)$. Obviously,

$$\begin{aligned} T(a) &= \sum_{x,y=1}^{p} \delta_p(x^2 - y^2 + a) \\ &= \frac{1}{p} \sum_{z=0}^{p-1} e^{2\pi i \frac{az}{p}} \sum_{x,y=1}^{p} e^{2\pi i \frac{z(x^2-y^2)}{p}} \\ &= p + \frac{1}{p} \sum_{z=1}^{p-1} e^{2\pi i \frac{az}{p}} \left| \sum_{x=1}^{p} e^{2\pi i \frac{zx^2}{p}} \right|^2. \end{aligned}$$

Using the fact that the modulus of the Gaussian sum equals $\sqrt{p}$, we obtain

$$T(a) = p + \sum_{z=1}^{p-1} e^{2\pi i \frac{az}{p}} = p + p\delta_p(a) - 1,$$

and, therefore, by (129)

$$\sum_{x=1}^{p} \left(\frac{x^2 + a}{p}\right) = p\delta_p(a) - 1. \tag{130}$$

Let $a_2 \not\equiv 0 \pmod{p}$. Then observing that

$$\left(\frac{a_2}{p}\right)\left(\frac{4\,a_2}{p}\right) = 1,$$

we get

$$\begin{aligned}\sigma_2 &= \left(\frac{a_2}{p}\right)\sum_{x=1}^{p}\left(\frac{4a_0a_2 + 4a_1a_2x + 4a_2^2x^2}{p}\right)\\ &= \left(\frac{a_2}{p}\right)\sum_{x=1}^{p}\left(\frac{(2a_2x + a_1)^2 + 4a_0a_2 - a_1^2}{p}\right)\\ &= \left(\frac{a_2}{p}\right)\sum_{x=1}^{p}\left(\frac{x^2 + 4a_0a_2 - a_1^2}{p}\right).\end{aligned}$$

Hence by (130) it follows that under $(a_2, p) = 1$

$$\sigma_2 = \sum_{x=1}^{p}\left(\frac{a_0 + a_1x + a_2x^2}{p}\right) = \left(\frac{a_2}{p}\right)\left[p\delta_p(4a_0a_2 - a_1^2) - 1\right]. \tag{131}$$

Note that, in particular,

$$\sum_{x=1}^{p}\left(\frac{x-a}{p}\right)\left(\frac{x-b}{p}\right) = p\delta_p(a-b) - 1. \tag{132}$$

Indeed, this equality follows at once from (131):

$$\begin{aligned}\sum_{x=1}^{p}\left(\frac{x-a}{p}\right)\left(\frac{x-b}{p}\right) &= \sum_{x=1}^{p}\left(\frac{ab - (a+b)x + x^2}{p}\right)\\ &= p\delta_p\left[4\,ab - (a+b)^2\right] - 1 = p\delta_p(a-b) - 1.\end{aligned}$$

Under $n \geqslant 3$ the investigation of the sums σ_n is much more complicated, except for some special cases.

Consider one of such special cases. Let $n \geqslant 3$ be odd, $p > n$ be a prime, $(a_1, p) = 1$, and

$$\sigma_n(a_1) = \sum_{x=1}^{p} \left(\frac{x^n + a_1 x}{p} \right).$$

Let us show that

$$\left| \sum_{x=1}^{p} \left(\frac{x^n + a_1 x}{p} \right) \right| \leqslant (n-1)\sqrt{p}. \tag{133}$$

In fact, since under $z \not\equiv 0 \pmod p$ the linear function zx runs through a complete residue system modulo p when x runs through a complete residue system modulo p, then

$$\sigma_n(a_1) = \sum_{x=1}^{p} \left(\frac{z^n x^n + a_1 z x}{p} \right).$$

Therefore,

$$\left|\sigma_n\left(a_1 z^{n-1}\right)\right| = \left| \sum_{x=1}^{p} \left(\frac{z^n x^n + a_1 z^n x}{p} \right) \right| = \left| \sum_{x=1}^{p} \left(\frac{x^n + a_1 x}{p} \right) \right| = |\sigma_n(a_1)|.$$

Squaring this equality and summing over z, we obtain

$$(p-1)\,|\sigma_n(a_1)|^2 = \sum_{z=1}^{p-1} \left|\sigma_n\left(a_1 z^{n-1}\right)\right|^2 = \sum_{\lambda=1}^{p-1} t(\lambda)\,|\sigma_n(\lambda)|^2, \tag{134}$$

where $t(\lambda)$ is the number of solutions of the congruence

$$a_1 z^{n-1} \equiv \lambda \pmod p.$$

Since $t(\lambda) \leqslant n-1$, then from (134) it follows that

$$\begin{aligned}
(p-1)\,|\sigma_n(a_1)|^2 &\leqslant (n-1) \sum_{\lambda=1}^{p-1} |\sigma_n(\lambda)|^2 \\
&= (n-1) \sum_{\lambda=1}^{p-1} \sum_{x,y=1}^{p} \left(\frac{x^n + \lambda x}{p} \right) \left(\frac{y^n + \lambda y}{p} \right) \\
&= (n-1) \sum_{x,y=1}^{p-1} \left(\frac{xy}{p} \right) \sum_{\lambda=1}^{p} \left(\frac{\lambda + x^{n-1}}{p} \right) \left(\frac{\lambda + y^{n-1}}{p} \right).
\end{aligned}$$

Hence, using the equality (132), we get the estimate (133):

$$\begin{aligned}
|\sigma_n(a_1)|^2 &\leqslant \frac{n-1}{p-1} \sum_{x,\,y=1}^{p-1} \left(\frac{xy}{p} \right) \left[p\delta_p\left(x^{n-1} - y^{n-1}\right) - 1 \right] \\
&= \frac{(n-1)p}{p-1} \sum_{x,\,y=1}^{p-1} \left(\frac{xy}{p} \right) \delta_p\left(x^{n-1} - y^{n-1}\right) \leqslant (n-1)^2 p, \\
|\sigma_n(a_1)| &\leqslant (n-1)\sqrt{p}.
\end{aligned}$$

Under odd $n \geqslant 3$ and $(a_n, p) = 1$ the same estimate holds for the general case also:

$$\left|\sum_{x=1}^{p}\left(\frac{a_0 + a_1 x + \ldots + a_n x^n}{p}\right)\right| \leqslant (n-1)\sqrt{p}. \tag{135}$$

For $n = 3$ this estimate was obtained by Hasse [13], under an arbitrary n it follows from more general results of A. Weil [48]. One can acquaint with elementary methods for obtaining estimates of the sums (135) by papers [35], [42], and [31].

As it was shown above, sums of Legendre's symbols for polynomials of the second degree can be evaluated with the help of Gaussian sums. Let us show that Gaussian sums can be used in estimating the simplest incomplete sums of Legendre's symbols:

$$\sigma(P) = \sum_{x=1}^{P}\left(\frac{x}{p}\right) \qquad (P < p).$$

Indeed, in the same way as in the estimation of incomplete exponential sums (see Theorem 2), we obtain

$$\sigma(P) = \sum_{x=1}^{p-1}\left(\frac{x}{p}\right)\sum_{y=1}^{P}\delta_p(x-y) = \frac{1}{p}\sum_{x=1}^{p-1}\left(\frac{x}{p}\right)\sum_{y=1}^{P}\sum_{z=0}^{p-1} e^{2\pi i\frac{z(x-y)}{p}}.$$

Hence, after interchanging the order of summation and singling out the summand with $z = 0$, by (41) and (42) it follows that

$$\sigma(P) = \frac{1}{p}\sum_{z=1}^{p-1}\left(\sum_{y=1}^{P} e^{-2\pi i\frac{zy}{p}}\right)\left(\sum_{x=1}^{p-1}\left(\frac{x}{p}\right)e^{2\pi i\frac{zx}{p}}\right),$$

$$|\sigma(P)| \leqslant \frac{1}{p}\sum_{z=1}^{p-1}\left|\sum_{y=1}^{P} e^{2\pi i\frac{zy}{p}}\right|\left|\sum_{x=1}^{p-1}\left(\frac{x}{p}\right)e^{2\pi i\frac{zx}{p}}\right|$$

$$\leqslant \frac{1}{\sqrt{p}}\sum_{z=1}^{p-1}\min\left(P, \frac{1}{2\left\|\frac{z}{p}\right\|}\right) \leqslant \sqrt{p}\log p.$$

Thus we obtain the estimate

$$\left|\sum_{x=1}^{P}\left(\frac{x}{p}\right)\right| \leqslant \sqrt{p}\log p. \tag{136}$$

Plainly, under $P > \sqrt{p}\log p$ this estimate is better than the trivial one.

The availability of a nontrivial estimate

$$\left|\sum_{x=1}^{P}\left(\frac{x}{p}\right)\right| < P$$

signalizes that on the interval $[1, P]$ there is at least one quadratic nonresidue modulo p. Let P_0 denote the least quadratic nonresidue. From (136) it follows that

$$P_0 \leqslant 1 + [\sqrt{p} \log p] = O(p^{\frac{1}{2}} \log p).$$

According to the conjecture enunciated by I. M. Vinogradov, under any $\varepsilon > 0$ the estimate

$$P_0 = O(p^{\varepsilon})$$

is valid, where the constant implied by the symbol "O" depends on ε only.

In this direction, the strongest result has the form

$$P_0 = O(p^{\gamma}),$$

where γ is any number greater than $\frac{1}{4\sqrt{e}}$. A proof of this result [4] is based upon the use of the Hasse–Weil estimate (135).

Let N_1 and N_2 be, respectively, the number of quadratic residues and quadratic nonresidues to modulus p among the first P positive integers. It is easy to obtain asymptotic formulas for N_1 and N_2 with the help of the estimate (136). Indeed, observing that the number of solutions of the congruence

$$y^2 \equiv x \pmod p, \qquad 1 \leqslant y \leqslant p,\ 1 \leqslant x \leqslant P,$$

under $P < p$ equals $2N_1$, we get

$$2N_1 = \sum_{x=1}^{P} \sum_{y=1}^{p} \delta_p(y^2 - x) = \sum_{x=1}^{P} \left[1 + \left(\frac{x}{p}\right)\right] = P + \theta\sqrt{p} \log p,$$

where $|\theta| \leqslant 1$ by (136). Hence, since $N_1 + N_2 = P$, we have

$$N_1 = \frac{1}{2} P + \frac{1}{2} \theta \sqrt{p} \log p,$$
$$N_2 = \frac{1}{2} P - \frac{1}{2} \theta \sqrt{p} \log p.$$

A question concerning the distribution of quadratic nonresidues in a sequence of values of recurrent functions is worked out just as easily.

Let $p > 2$ be a prime and $\psi(x)$ be a recurrent function of the n-th order ($n \geqslant 2$) with a period τ to modulus p. Let N_0, N_1, and N_2 denote, respectively, the number of zeros to modulus p, quadratic residues and quadratic nonresidues in a period of the function $\psi(x)$. Obviously,

$$N_0 = \sum_{x=1}^{\tau} \delta_p[\psi(x)] = \frac{1}{p} \sum_{x=1}^{\tau} \sum_{z=0}^{p-1} e^{2\pi i \frac{z\psi(x)}{p}} = \frac{\tau}{p} + \frac{1}{p} \sum_{z=1}^{p-1} \sum_{x=1}^{\tau} e^{2\pi i \frac{z\psi(x)}{p}}.$$

Hence, using Theorem 13, we get

$$\left|N_0 - \frac{\tau}{p}\right| \leqslant \frac{1}{p}\sum_{z=1}^{p-1}\left|\sum_{x=1}^{\tau} e^{2\pi i \frac{z\psi(x)}{p}}\right| \leqslant \left(1 - \frac{1}{p}\right)p^{\frac{n}{2}},$$

$$N_0 = \frac{\tau}{p} + \theta_0\left(1 - \frac{1}{p}\right)p^{\frac{n}{2}}, \tag{137}$$

where $|\theta_0| \leqslant 1$. In the same way, using Theorems 3 and 13, we obtain

$$N_0 + 2N_1 = \sum_{x=1}^{\tau}\sum_{y=1}^{p}\delta_p\left[\psi(x) - y^2\right] = \frac{1}{p}\sum_{x=1}^{\tau}\sum_{y=1}^{p}\sum_{z=0}^{p-1} e^{2\pi i \frac{z\left[\psi(x)-y^2\right]}{p}}$$

$$= \tau + \frac{1}{p}\sum_{z=1}^{p-1}\left(\sum_{y=1}^{p} e^{-2\pi i \frac{zy^2}{p}}\right)\left(\sum_{x=1}^{\tau} e^{2\pi i \frac{z\psi(x)}{p}}\right).$$

Therefore,

$$|N_0 + 2N_1 - \tau| \leqslant \frac{1}{p}\sum_{z=1}^{p-1}\left|\sum_{y=1}^{p} e^{2\pi i \frac{zy^2}{p}}\right|\left|\sum_{x=1}^{\tau} e^{2\pi i \frac{z\psi(x)}{p}}\right|$$

$$\leqslant \left(1 - \frac{1}{p}\right)p^{\frac{n+1}{2}},$$

$$N_0 + 2N_1 = \tau + \theta_1\left(1 - \frac{1}{p}\right)p^{\frac{n+1}{2}}, \qquad |\theta_1| \leqslant 1. \tag{138}$$

Now, observing that $N_0 + N_1 + N_2 = \tau$, from (137) and (138) we get the asymptotic formulas for N_1 and N_2:

$$N_1 = \frac{p-1}{2p}\tau + \frac{p-1}{2p}p^{\frac{n}{2}}(\theta_1\sqrt{p} - \theta_0),$$

$$N_2 = \frac{p-1}{2p}\tau - \frac{p-1}{2p}p^{\frac{n}{2}}(\theta_1\sqrt{p} + \theta_0).$$

If the period of the recurrent function is sufficiently large,

$$\tau > p^{\frac{n}{2}}(\sqrt{p} + 1), \tag{139}$$

then the magnitudes N_1 and N_2 will be positive and, therefore, both quadratic residues and quadratic nonresidues to modulus p will occur among terms of the recurrent sequence. Note that for sequences of the third order the condition (139) is nearly of the best possible kind. By (139) recurrent sequences of the third order, whose period is greater than $p^2 + p\sqrt{p}$, contain quadratic nonresidues to modulus p.

We shall show that there exist sequences of the third order with period $\frac{1}{2}(p^2 - p)$, which do not contain quadratic nonresidues.

Indeed, let g be a primitive root to modulus p. Consider the function $\psi(x)$ satisfying the equation of the third order

$$\psi(x) = 3g^2\psi(x-1) - 3g^4\psi(x-2) + g^6\psi(x-3)$$

and determined by initial conditions

$$\psi(1) = g^2, \qquad \psi(2) = 4g^4, \qquad \psi(3) = 9g^6.$$

It is easy to verify that $\psi(x) = x^2g^{2x}$. Obviously, the sequence of values of this function does not contain quadratic nonresidues. Let τ denote the least period of the function $\psi(x)$ to modulus p. Then the congruence

$$(x+\tau)^2g^{2x+2\tau} \equiv x^2g^{2x} \pmod p$$

holds for any integer x. Hence we get without difficulty that $\tau = \frac{1}{2}(p^2 - p)$.

Thus, under $n = 3$ the bound (139) for the magnitude of periods of recurrent functions, in whose values quadratic nonresidues occur, has the precise order and the constant in it cannot be improved more than twice.

CHAPTER II

WEYL'S SUMS

§ 11. Weyl's method

In Chapter I, the Weyl sums of the first degree were considered and it was shown that the estimate

$$\left|\sum_{x=1}^{P} e^{2\pi i\,\alpha x}\right| \leqslant \min\left(P, \frac{1}{2\|\alpha\|}\right) \tag{140}$$

holds for them. The basic idea of Weyl's method consists in reducing the estimation of a sum of an arbitrary degree $n \geqslant 2$

$$S(P) = \sum_{x=1}^{P} e^{2\pi i\,(\alpha_1 x + \ldots + \alpha_n x^n)}$$

to the estimation of a sum of degree $n-1$ and, ultimately, to the use of the estimate (140). We have already met the reduction of the degree of an exponential sum in proving the theorem on the modulus of the Gauss sum. In the Gauss theorem, the square of the modulus of the exponential sum of the second degree was transformed with the help of linear change of variable in summation into a double sum, in which one of summations was reduced to the evaluation of a sum of the first degree. Similar but technically more complicated considerations are used for the reduction of the degree of sums in the Weyl method as well.

In deducing estimates of Weyl's sums we shall need the following inequalities:

$$\left(\sum_{x=1}^{P} u_x v_x\right)^k \leqslant \left(\sum_{x=1}^{P} u_x\right)^{k-1} \sum_{x=1}^{P} u_x v_x^k, \tag{141}$$

$$\left(\sum_{x=1}^{P} v_x\right)^k \leqslant P^{k-1} \sum_{x=1}^{P} v_x^k, \tag{142}$$

$$\left(\sum_{x=1}^{P} u_x v_x\right)^2 \leqslant \sum_{x=1}^{P} u_x^2 \sum_{x=1}^{P} v_x^2. \tag{143}$$

These inequalities hold under $u_x \geqslant 0$, $v_x \geqslant 0$, and an arbitrary positive integer k.

Let us prove the inequality (141). Denote by σ_k the sums

$$\sigma_0 = \sum_{x=1}^{P} u_x, \qquad \sigma_k = \sum_{x=1}^{P} u_x v_x^k \qquad (k = 1, 2, \ldots).$$

If $\sigma_0 = 0$ or $\sigma_0 \neq 0$ and $k = 1$, then the inequality (141) is trivial. We shall assume that $\sigma_0 \neq 0$ and $k \geqslant 2$. Since

$$v_x^k - v_y v_x^{k-1} + v_y^k - v_x v_y^{k-1} = (v_x - v_y)\left(v_x^{k-1} - v_y^{k-1}\right) \geqslant 0,$$

then, obviously,

$$0 \leqslant \sum_{x,y=1}^{P} u_x u_y \left(v_x^k - v_y v_x^{k-1} + v_y^k - v_x v_y^{k-1}\right) = 2(\sigma_0 \sigma_k - \sigma_1 \sigma_{k-1}).$$

But then

$$\sigma_k \geqslant \frac{\sigma_1}{\sigma_0} \sigma_{k-1} \geqslant \frac{\sigma_1^2}{\sigma_0^2} \sigma_{k-2} \geqslant \ldots \geqslant \frac{\sigma_1^k}{\sigma_0^{k-1}},$$

and, therefore,

$$\sigma_1^k \leqslant \sigma_0^{k-1} \sigma_k.$$

The last inequality coincides with (141).

The inequality (142) is obtained from (141) under $u_1 = \ldots = u_P = 1$. The inequality (143) follows from (141) also. Indeed, denote by $\sum^*$ the sum extended over those values x, for which $u_x \neq 0$. Then, setting $k = 2$ in (141), we obtain the inequality (143):

$$\left(\sum_{x=1}^{P} u_x v_x\right)^2 = \left(\sum_{x=1}^{P}{}^{*} u_x^2 \left(u_x^{-1} v_x\right)\right)^2 \leqslant \sum_{x=1}^{P}{}^{*} u_x^2 \sum_{x=1}^{P}{}^{*} u_x^2 \left(u_x^{-1} v_x\right)^2 \leqslant \sum_{x=1}^{P} u_x^2 \sum_{x=1}^{P} v_x^2.$$

Let $y_1 \geqslant 0$ be an integer. We shall use $\underset{y_1}{\Delta} f(x)$ to denote a finite difference of a function $f(x)$:

$$\underset{y_1}{\Delta} f(x) = f(x + y_1) - f(x).$$

Under $k \geqslant 1$ we determine a finite difference of the k-th order $\underset{y_1, \ldots, y_k}{\Delta} f(x)$ with the help of the equality

$$\underset{y_1, \ldots, y_k}{\Delta} f(x) = \underset{y_k}{\Delta} \left[\underset{y_1, \ldots, y_{k-1}}{\Delta} f(x) \right].$$

It is easily seen that $\underset{y_1,\dots,y_k}{\Delta} f(x)$ does not depend on the order of the arrangement of quantities $y_1, \dots, y_k$. So, for instance,

$$\underset{y_1,y_2}{\Delta} f(x) = \underset{y_2}{\Delta}\left[\underset{y_1}{\Delta} f(x)\right] = \underset{y_2}{\Delta}[f(x+y_1) - f(x)]$$
$$= f(x+y_1+y_2) - f(x+y_2) - f(x+y_1) + f(x) = \underset{y_2,y_1}{\Delta} f(x).$$

Let $f(x)$ be a polynomial of degree $n \geqslant 2$:

$$f(x) = \alpha_0 + \alpha_1 x + \ldots + \alpha_n x^n.$$

We shall show that for its finite difference of the order $n-1$ the equality

$$\underset{y_1,\dots,y_{n-1}}{\Delta} f(x) = n!\,\alpha_n y_1 \ldots y_{n-1} x + \beta, \tag{144}$$

where β depends only on coefficients of the polynomial $f(x)$ and on quantities $y_1, \dots, y_{n-1}$, is valid.

Indeed, for polynomials of the second degree $f_2(x) = \alpha_2 x^2 + \alpha_1 x + \alpha_0$ this equality follows immediately from the definition of a finite difference:

$$\underset{y_1}{\Delta} f_2(x) = \alpha_2(x+y_1)^2 + \alpha_1(x+y_1) + \alpha_0 - (\alpha_2 x^2 + \alpha_1 x + \alpha_0)$$
$$= 2\alpha_2 y_1 x + \alpha_2 y_1^2 + \alpha_1 y_1 = 2\alpha_2 y_1 x + \beta_2.$$

Apply induction. Let under a certain $k \geqslant 2$ the equality

$$\underset{y_1,\dots,y_{k-1}}{\Delta} f_k(x) = k!\,\alpha_k y_1 \ldots y_{k-1} x + \beta_k \tag{145}$$

be valid for every polynomial $f_k(x) = \alpha_k x^k + \ldots + \alpha_0$. Then for $f_{k+1}(x) = \alpha_{k+1}x^{k+1} + \ldots + \alpha_0$ we get

$$\underset{y_1,\dots,y_k}{\Delta} f_{k+1}(x) = \underset{y_1,\dots,y_{k-1}}{\Delta} [f_{k+1}(x+y_k) - f_{k+1}(x)]$$
$$= \underset{y_1,\dots,y_{k-1}}{\Delta} [(k+1)\alpha_{k+1} y_k x^k + \ldots]$$
$$= (k+1)!\,\alpha_{k+1} y_1 \ldots y_k x + \beta_{k+1},$$

by that the equality (145) is proved for any $k \geqslant 2$. In particular, under $k = n$ we obtain the equality (144). The following lemma is central in Weyl's method.

LEMMA 12. *Under any $k \geqslant 1$ we have*

$$\left|\sum_{x=1}^{P} e^{2\pi i f(x)}\right|^{2^k} \leqslant 2^{2^k-1} P^{2^k-(k+1)} \sum_{y_1=0}^{P_1-1} \ldots \sum_{y_k=0}^{P_k-1} \left|\sum_{x=1}^{P_{k+1}} e^{2\pi i \underset{y_1,\dots,y_k}{\Delta} f(x)}\right|,$$

where $P_1 = P$ and under $\nu = 1, 2, \dots, k$ quantities $P_{\nu+1}$ are determined by the equality $P_{\nu+1} = P_\nu - y_\nu$.

Proof. At first we shall show that the assertion of the lemma holds under $k = 1$. Indeed,

$$\left|\sum_{x=1}^{P_1} e^{2\pi i\, f(x)}\right|^2 = \sum_{x,\, y=1}^{P_1} e^{2\pi i\,[f(y)-f(x)]}$$
$$= P_1 + \sum_{x<y} e^{2\pi i\,[f(y)-f(x)]} + \sum_{x>y} e^{2\pi i\,[f(y)-f(x)]}$$
$$\leqslant P_1 + 2\left|\sum_{x=1}^{P_1-1}\sum_{y=1}^{P_1-x} e^{2\pi i\,[f(x+y)-f(x)]}\right|.$$

Hence, after interchanging the order of summation, it follows that

$$\left|\sum_{x=1}^{P_1} e^{2\pi i\, f(x)}\right|^2 \leqslant P_1 + 2\sum_{y_1=1}^{P_1-1}\left|\sum_{x=1}^{P_1-y_1} e^{2\pi i \underset{y_1}{\Delta} f(x)}\right| \leqslant 2\sum_{y_1=0}^{P_1-1}\left|\sum_{x=1}^{P_2} e^{2\pi i \underset{y_1}{\Delta} f(x)}\right|.$$

Raise this inequality to the power 2^{k-1}. Then according to (142) we obtain

$$\left|\sum_{x=1}^{P_1} e^{2\pi i\, f(x)}\right|^{2^k} \leqslant 2^{2^{k-1}} P_1^{2^{k-1}-1} \sum_{y_1=0}^{P_1-1}\left|\sum_{x=1}^{P_2} e^{2\pi i \underset{y_1}{\Delta} f(x)}\right|^{2^{k-1}}. \tag{146}$$

Applying the inequality (146) to its right-hand side in succession and observing that $P_1 = P$ and $P_\nu \leqslant P$, we arrive at the assertion of the lemma:

$$\left|\sum_{x=1}^{P_1} e^{2\pi i\, f(x)}\right|^{2^k} \leqslant \left(\prod_{\nu=1}^{k} 2^{2^{k-\nu}} P_\nu^{2^{k-\nu}-1}\right)\sum_{y_1=0}^{P_1-1}\cdots\sum_{y_k=0}^{P_k-1}\left|\sum_{x=1}^{P_{k+1}} e^{2\pi i \underset{y_1,\ldots,y_k}{\Delta} f(x)}\right|$$
$$\leqslant 2^{2^k-1} P^{2^k-(k+1)} \sum_{y_1=0}^{P_1-1}\cdots\sum_{y_k=0}^{P_k-1}\left|\sum_{x=1}^{P_{k+1}} e^{2\pi i \underset{y_1,\ldots,y_k}{\Delta} f(x)}\right|.$$

LEMMA 13. *Let λ and $x_1, \ldots, x_n$ be positive integers. Denote by $\tau_n(\lambda)$ the number of solutions of the equation $x_1 \ldots x_n = \lambda$. Then under any ε $(0 < \varepsilon \leqslant 1)$ we have*

$$\tau_n(\lambda) \leqslant C_n(\varepsilon)\lambda^\varepsilon,$$

where the constant $C_n(\varepsilon)$ depends on n and ε only.

Proof. Let $\alpha \geqslant 1$, $p \geqslant 2$, and $0 < \varepsilon \leqslant 1$. Since

$$1 + \alpha\varepsilon \log p < e^{\alpha\varepsilon \log p} = p^{\alpha\varepsilon},$$

then for any $p \geqslant 2$

$$1+\alpha < \frac{1}{\varepsilon \log 2}(1+\alpha\varepsilon \log p) < \frac{1}{\varepsilon \log 2} p^{\alpha\varepsilon}. \tag{147}$$

If $p \geqslant e^{\frac{1}{\varepsilon}}$, then the coefficient $\frac{1}{\varepsilon \log 2}$ in this estimate may be omitted:

$$1+\alpha < e^{\alpha} = \left(e^{\frac{1}{\varepsilon}}\right)^{\alpha\varepsilon} \leqslant p^{\alpha\varepsilon}. \tag{148}$$

Under $\lambda = 1$ the assertion of the lemma is obvious. Let $\lambda \geqslant 2$ be given by its prime factorization:

$$\lambda = p_1^{\alpha_1} \dots p_s^{\alpha_s} \qquad (p_1 < p_2 < \dots < p_s).$$

Estimate the number of divisors of λ. Let $p_r < e^{\frac{1}{\varepsilon}} \leqslant p_{r+1}$. Then, applying the estimate (147) for $p_1, \dots, p_r$ and using the estimate (148) for $p_{r+1}, \dots, p_s$, we obtain

$$\tau(\lambda) = (1+\alpha_1)\dots(1+\alpha_s) \leqslant \left(\frac{1}{\varepsilon \log 2}\right)^r p_1^{\alpha_1\varepsilon} \dots p_s^{\alpha_s\varepsilon} = \left(\frac{1}{\varepsilon \log 2}\right)^r \lambda^{\varepsilon}.$$

Hence, because the number of primes, which are less than $e^{\frac{1}{\varepsilon}}$, does not exceed $e^{\frac{1}{\varepsilon}}$, it follows that

$$\tau(\lambda) \leqslant \left(\frac{1}{\varepsilon \log 2}\right)^{e^{\frac{1}{\varepsilon}}} \lambda^{\varepsilon} = C(\varepsilon)\lambda^{\varepsilon}.$$

Replace ε by $\frac{\varepsilon}{n}$ in this estimate. Then, observing that $\tau_n(\lambda) \leqslant [\tau(\lambda)]^n$, we get the assertion of the lemma:

$$\tau_n(\lambda) \leqslant \left[C\left(\frac{\varepsilon}{n}\right)\lambda^{\frac{\varepsilon}{n}}\right]^n = C_n(\varepsilon)\lambda^{\varepsilon}.$$

LEMMA 14. *Let $P \geqslant 2$ and*

$$\alpha = \frac{a}{q} + \frac{\theta}{q^2}, \qquad (a, q) = 1, \qquad |\theta| \leqslant 1.$$

Then under any positive integer Q and an arbitrary real β we have

$$1^{\circ}. \qquad \sum_{x=1}^{Q} \min\left(P, \frac{1}{\|\alpha x + \beta\|}\right) \leqslant 4\left(1+\frac{Q}{q}\right)(P + q \log P),$$

$$2^{\circ}. \qquad \sum_{x=1}^{Q} \min\left(P^2, \frac{1}{\|\alpha x + \beta\|^2}\right) \leqslant 4P\left(1+\frac{Q}{q}\right)(P + q).$$

Proof. Let us represent β in the form

$$\beta = \frac{b}{q} + \frac{\theta_1}{q},$$

where b is an integer, $|\theta_1| < 1$ and the sign of θ_1 is opposite to the sign of θ. Then under $1 \leqslant x \leqslant q$ we obtain

$$\alpha x + \beta = \frac{ax+b}{q} + \frac{\theta x}{q^2} + \frac{\theta_1}{q},$$

$$\left\| \frac{ax+b}{q} \right\| = \left\| \alpha x + \beta - \left(\frac{\theta x}{q^2} + \frac{\theta_1}{q} \right) \right\| \leqslant \|\alpha x + \beta\| + \frac{1}{q}. \tag{149}$$

At first we shall show that

$$\sum_{x=1}^{q} \min\left(P, \frac{1}{\|\alpha x + \beta\|}\right) \leqslant 3P + 4q \log P. \tag{150}$$

Indeed, if q or P is less than four, then this estimate is trivial. Let $q \geqslant 4$ and $P \geqslant 4$. Then, according to (149) for those values x, under which

$$\left\| \frac{ax+b}{q} \right\| \geqslant \frac{2}{q},$$

we have

$$\|\alpha x + \beta\| \geqslant \left\| \frac{ax+b}{q} \right\| - \frac{1}{q} \geqslant \frac{1}{2} \left\| \frac{ax+b}{q} \right\|.$$

This estimate may be used for all x within the interval $1 \leqslant x \leqslant q$ with the exception of those, for which

$$ax + b \equiv 0, \pm 1 \pmod q.$$

Since $(a, q) = 1$, then $ax + b$ runs through a complete residue system modulo q, when x runs through a complete residue system modulo p. Therefore

$$\begin{aligned} \sum_{x=1}^{q} \min\left(P, \frac{1}{\|\alpha x + \beta\|}\right) &\leqslant 3P + \sum_{ax+b \not\equiv 0, \pm 1} \min\left(P, \frac{2}{\left\|\frac{ax+b}{q}\right\|}\right) \\ &= 3P + \sum_{2 \leqslant x < q} \min\left(P, \frac{2}{\left\|\frac{x}{q}\right\|}\right) \\ &\leqslant 3P + 2 \sum_{2 \leqslant x \leqslant \frac{q}{2}} \min\left(P, \frac{2q}{x}\right). \end{aligned} \tag{151}$$

Hence, observing that

$$\sum_{2\leqslant x\leqslant\frac{q}{2}} \min\left(P, \frac{2q}{x}\right) \leqslant \sum_{2\leqslant x<\frac{2q}{P}+1} P + 2q \sum_{1+\frac{2q}{P}\leqslant x\leqslant\frac{q}{2}} \frac{1}{x}$$

$$\leqslant 2q + 2q \int_{\frac{2q}{P}}^{\frac{q}{2}} \frac{dx}{x} \leqslant 2q \log P,$$

we get the estimate (150).

Now choose $Q_1 = 1+\left[\frac{1}{q}\,Q\right]$ and replace the quantity x by qx_1+x_2 in the estimate 1°. Then, using the inequality (150), we obtain

$$\sum_{x=1}^{Q} \min\left(P, \frac{1}{\|\alpha x+\beta\|}\right) \leqslant \sum_{x_1=0}^{Q_1-1} \sum_{x_2=1}^{q} \min\left(P, \frac{1}{\|\alpha x_2+\alpha q x_1+\beta\|}\right)$$

$$\leqslant Q_1(3P+4q\log P).$$

Hence, since $Q_1 < 1+\frac{Q}{q}$, the first assertion of the lemma follows.

To prove the assertion 2°, at first we will show that

$$\sum_{x=1}^{q} \min\left(P^2, \frac{1}{\|\alpha x+\beta\|^2}\right) \leqslant 3P^2+4Pq. \tag{152}$$

In fact, as in the proof of the estimate (151), we get

$$\sum_{x=1}^{q} \min\left(P^2, \frac{1}{\|\alpha x+\beta\|^2}\right) \leqslant 3P^2 + 2 \sum_{2\leqslant x\leqslant\frac{q}{2}} \min\left(P^2, \frac{4q^2}{x^2}\right).$$

Hence, because of

$$\sum_{2\leqslant x\leqslant\frac{q}{2}} \min\left(P^2, \frac{4q^2}{x^2}\right) \leqslant \sum_{2\leqslant x<\frac{2q}{P}+1} P^2 + 4q^2 \sum_{x\geqslant\frac{2q}{P}+1} \frac{1}{x^2}$$

$$\leqslant 2Pq + 4q^2 \int_{\frac{2q}{P}}^{\infty} \frac{dx}{x^2} = 4Pq,$$

we obtain the estimate (152). Therefore,

$$\sum_{x=1}^{Q}\min\left(P^2,\ \frac{1}{\|\alpha x+\beta\|^2}\right)\leqslant\sum_{x_1=1}^{Q_1}\sum_{x_2=1}^{q}\min\left(P^2,\ \frac{1}{\|\alpha x_2+\alpha q x_1+\beta\|^2}\right)$$
$$\leqslant Q_1(3P^2+4Pq)\leqslant 4P\left(1+\frac{Q}{q}\right)(P+q),$$

thus completing the proof of the lemma.

Note. Let m be an arbitrary positive integer and

$$\alpha=\frac{a}{q}+\frac{\theta}{q^2},\qquad (a,\ q)=1,\quad |\theta|\leqslant 1.$$

Then under any β and any ε $(0<\varepsilon\leqslant 1)$ we have

$$\sum_{|x|<Q}\min\left(P,\ \frac{1}{\|\alpha m x+\beta\|}\right)\leqslant\frac{8m}{\varepsilon}\left(1+\frac{Q}{q}\right)(P+q)P^{\varepsilon}. \tag{153}$$

Really, under $P<3$ this estimation is trivial. Let $P\geqslant 3$. Then $1<\log P<\frac{1}{\varepsilon}P^{\varepsilon}$ and

$$\sum_{|x|<Q}\min\left(P,\ \frac{1}{\|\alpha m x+\beta\|}\right)\leqslant\sum_{|x|<mQ}\min\left(P,\ \frac{1}{\|\alpha x+\beta\|}\right)$$
$$=\sum_{x=1}^{2mQ-1}\min\left(P,\ \frac{1}{\|\alpha x+\beta-\alpha m Q\|}\right)$$
$$\leqslant 4\left(1+\frac{2mQ-1}{q}\right)(P+q\log P).$$

Hence the estimate (153) follows obviously.

THEOREM 14. *Let $n\geqslant 2$, $f(x)=\alpha_1 x+\ldots+\alpha_n x^n$, and*

$$\alpha_n=\frac{a}{q}+\frac{\theta}{q^2},\qquad (a,\ q)=1,\quad |\theta|\leqslant 1.$$

If $P\leqslant q\leqslant P^{n-1}$, then under any positive $\varepsilon<1$ we have

$$\left|\sum_{x=1}^{P}e^{2\pi i f(x)}\right|\leqslant C(n,\ \varepsilon)P^{1-\frac{1-\varepsilon}{2^{n-1}}},$$

where the constant $C(n,\ \varepsilon)$ does not depend on P.

Proof. It follows from Lemma 12 under $k = n-1$ that

$$\left|\sum_{x=1}^{P} e^{2\pi i\, f(x)}\right|^{2^{n-1}} \leqslant 2^{2^{n-1}} P^{2^{n-1}-n} \sum_{y_1=0}^{P_1-1} \cdots \sum_{y_{n-1}=0}^{P_{n-1}-1} \left|\sum_{x=1}^{P_n} e^{2\pi i \underset{y_1,\dots,y_{n-1}}{\Delta} f(x)}\right|, \tag{154}$$

where $P_1 = P$ and $P_{\nu+1} = P_\nu - y_\nu$ $(\nu = 1, 2, \dots, n-1)$. Since by (144)

$$\underset{y_1,\dots,y_{n-1}}{\Delta} f(x) = \alpha_n n!\, y_1 \dots y_{n-1} x + \beta,$$

then using the estimate (11), we obtain

$$\left|\sum_{x=1}^{P_n} e^{2\pi i \underset{y_1,\dots,y_{n-1}}{\Delta} f(x)}\right| = \left|\sum_{x=1}^{P_n} e^{2\pi i\, \alpha_n n!\, y_1 \dots y_{n-1} x}\right|$$
$$\leqslant \min\left(P_n, \frac{1}{2\,\|\alpha_n n!\, y_1 \dots y_{n-1}\|}\right).$$

Substituting this estimate into (154) and singling out the summands, in which quantities y_ν being equal to zero occur, we get

$$\left|\sum_{x=1}^{P} e^{2\pi i\, f(x)}\right|^{2^{n-1}} \leqslant n 2^{2^{n-1}} P^{2^{n-1}-1}$$
$$+ 2^{2^{n-1}} P^{2^{n-1}-n} \sum_{y_1,\dots,y_{n-1}=1}^{P} \min\left(P, \frac{1}{\|n!\, y_1 \dots y_{n-1}\alpha_n\|}\right).$$

Collect summands with the same values of the product $y_1 \dots y_{n-1}$. Since by Lemma 13 for the number of solutions of the equation $y_1 \dots y_{n-1} = \lambda$ we have the estimate $\tau_{n-1}(\lambda) \leqslant C_{n-1}(\varepsilon)\lambda^\varepsilon$ under any positive $\varepsilon \leqslant 1$, then

$$\sum_{y_1,\dots,y_{n-1}=1}^{P} \min\left(P, \frac{1}{\|n!\, y_1 \dots y_{n-1}\alpha_n\|}\right)$$
$$\leqslant \sum_{\lambda=1}^{P^{n-1}} \tau_{n-1}(\lambda) \min\left(P, \frac{1}{\|n!\, \lambda\alpha_n\|}\right)$$
$$\leqslant C_{n-1}(\varepsilon) P^{(n-1)\varepsilon} \sum_{\lambda=1}^{P^{n-1}} \min\left(P, \frac{1}{\|n!\, \lambda\alpha_n\|}\right). \tag{155}$$

Now, using the condition $P \leqslant q \leqslant P^{n-1}$ and the note of Lemma 14, we obtain

$$\sum_{y_1,\dots,y_{n-1}=1}^{P} \min\left(P, \frac{1}{\|n!\, y_1 \dots y_{n-1}\alpha_n\|}\right)$$
$$\leqslant 8\,n!\, \frac{C_{n-1}(\varepsilon)}{\varepsilon} \left(1 + \frac{P^{n-1}}{q}\right)(P+q)P^{\varepsilon n}$$
$$\leqslant 32\,n!\, \frac{C_{n-1}(\varepsilon)}{\varepsilon} P^{n-1+\varepsilon n},$$

and, therefore,

$$\left|\sum_{x=1}^{P} e^{2\pi i\, f(x)}\right|^{2^{n-1}} \leqslant n\, 2^{2^{n-1}} P^{2^{n-1}-1} + 32\, n!\, 2^{2^{n-1}} \frac{C_{n-1}(\varepsilon)}{\varepsilon} P^{2^{n-1}-1+\varepsilon n}.$$

Hence after replacing ε by $\frac{\varepsilon}{n}$ we get the theorem assertion:

$$\left|\sum_{x=1}^{P} e^{2\pi i\, f(x)}\right| \leqslant \left(64\, n!\, 2^{2^{n-1}} \frac{nC_{n-1}(\frac{\varepsilon}{n})}{\varepsilon}\right)^{\frac{1}{2^{n-1}}} P^{1-\frac{1-\varepsilon}{2^{n-1}}}$$
$$= C(n,\, \varepsilon)\, P^{1-\frac{1-\varepsilon}{2^{n-1}}}.$$

Note. A nontrivial (with respect to the order) estimate of Weyl's sum of degree $n \geqslant 2$ can be obtained not only on the interval $P \leqslant q \leqslant P^{n-1}$ indicated in Theorem 14, but also on the interval

$$P^{\varepsilon_1} \leqslant q \leqslant P^{n-\varepsilon_1} \qquad (0 < \varepsilon_1 < 1) \tag{156}$$

containing the former within it.

Indeed, in this case by (155) we have the estimate

$$\left|\sum_{x=1}^{P} e^{2\pi i\, f(x)}\right| \leqslant C(n, \varepsilon, \varepsilon_1) P^{1-\frac{\varepsilon_1-\varepsilon}{2^{n-1}}},$$

having nontrivial order under any positive $\varepsilon < \varepsilon_1$. The further extension of the interval (156) to $q \leqslant P^n$ is impossible, because under $q = P^n$ there exist Weyl's sums of degree n, for which

$$\left|\sum_{x=1}^{P} e^{2\pi i\, f(x)}\right| > \frac{1}{2} P.$$

In order to convince ourselves in it, let us choose, for example, $n \geqslant 12$, $q = P^n$, and

$$f(x) = \frac{x^{n+1} - (x-1)^{n+1}}{(n+1)P^n} = \frac{1}{q} x^n + \alpha_{n-1} x^{n-1} + \ldots + \alpha_0.$$

Then, since

$$\left|e^{2\pi i\, f(x)} - 1\right| = 2|\sin \pi f(x)| \leqslant 2\pi \frac{x^{n+1} - (x-1)^{n+1}}{(n+1)P^n},$$

so, obviously,

$$\sum_{x=1}^{P} \left| e^{2\pi i\, f(x)} - 1 \right| \leqslant \frac{2\pi}{(n+1)P^n} \sum_{x=1}^{P} \left[x^{n+1} - (x-1)^{n+1} \right] = \frac{2\pi}{n+1} P,$$

and, therefore,

$$\left| \sum_{x=1}^{P} e^{2\pi i\, f(x)} \right| = \left| P + \sum_{x=1}^{P} \left(e^{2\pi i\, f(x)} - 1 \right) \right|$$

$$\geqslant P - \sum_{x=1}^{P} \left| e^{2\pi i\, f(x)} - 1 \right| \geqslant \left(1 - \frac{2\pi}{n+1} \right) P > \frac{1}{2} P.$$

§ 12. Systems of equations

A method proposed by Mordell for the estimation of complete rational sums consists in the reduction of the estimation of an individual sum to the estimation of the mean value

$$\frac{1}{p^n} \sum_{a_1=1}^{p} \cdots \sum_{a_n=1}^{p} \left| \sum_{x=1}^{p} e^{2\pi i \frac{a_1 x + \ldots + a_n x^n}{p}} \right|^{2k}$$

under $k = n$ or, in other words, to the estimation of the number of solutions of the system of congruences

$$\left. \begin{array}{c} x_1 + \ldots + x_k \equiv y_1 + \ldots + y_k \\ \ldots\ldots\ldots\ldots\ldots\ldots\ldots\ldots \\ x_1^n + \ldots + x_k^n \equiv y_1^n + \ldots + y_k^n \end{array} \right\} \pmod p.$$

Similarly, in Vinogradov's method the estimation of Weyl's sum is reduced to the estimation, under a certain k, the mean value of the quantity

$$\left| \sum_{x=1}^{P} e^{2\pi i (\alpha_1 x + \ldots + \alpha_n x^n)} \right|^{2k}$$

being equal to

$$\int_0^1 \cdots \int_0^1 \left| \sum_{x=1}^{P} e^{2\pi i (\alpha_1 x + \ldots + \alpha_n x^n)} \right|^{2k} d\alpha_1 \ldots d\alpha_n$$

and, as it will be shown below, coinciding with the number of solutions of the system of equations

$$\left.\begin{array}{l} x_1 + \ldots + x_k = y_1 + \ldots + y_k \\ \cdots\cdots\cdots\cdots\cdots\cdots\cdots\cdots \\ x_1^n + \ldots + x_k^n = y_1^n + \ldots + y_k^n \end{array}\right\}, \qquad 1 \leqslant x_j, y_j \leqslant P. \tag{157}$$

Denote by $S(\alpha_1, \ldots, \alpha_n)$ the Weyl sum

$$S(\alpha_1, \ldots, \alpha_n) = \sum_{x=1}^{P} e^{2\pi i (\alpha_1 x + \ldots + \alpha_n x^n)}.$$

Let $n \geqslant 2$, $\lambda_1, \ldots, \lambda_n$ be fixed integers, and $N_{k,n}^{(P)}(\lambda_1, \ldots, \lambda_n)$ be the number of integral solutions of the system of equations

$$\left.\begin{array}{l} x_1 + \ldots + x_k - (y_1 + \ldots + y_k) = \lambda_1 \\ \cdots\cdots\cdots\cdots\cdots\cdots\cdots\cdots\cdots\cdots \\ x_1^n + \ldots + x_k^n - (y_1^n + \ldots + y_k^n) = \lambda_n \end{array}\right\}, \tag{158}$$

in which the quantities x_j and y_j vary within the limits

$$1 \leqslant x_j \leqslant P, \qquad 1 \leqslant y_j \leqslant P \qquad (j = 1, 2, \ldots, k).$$

Obviously, under $\lambda_1 = \ldots = \lambda_n = 0$ this system of equations coincides with the system (157).

Let us consider the simplest properties of such systems. First of all we shall show that under any positive integer k we have

$$|S(\alpha_1, \ldots, \alpha_n)|^{2k} = \sum_{\lambda_1, \ldots, \lambda_n} N_{k,n}^{(P)}(\lambda_1, \ldots, \lambda_n) e^{2\pi i (\alpha_1 \lambda_1 + \ldots + \alpha_n \lambda_n)}, \tag{159}$$

where the range of summation is

$$|\lambda_\nu| < kP^\nu \qquad (\nu = 1, 2, \ldots, n).$$

Indeed, since

$$S^k(\alpha_1, \ldots, \alpha_n) = \sum_{x_1, \ldots, x_k = 1}^{P} e^{2\pi i (\alpha_1 (x_1 + \ldots + x_k) + \ldots + \alpha_n (x_1^n + \ldots + x_k^n))},$$

then, obviously,

$$|S(\alpha_1, \ldots, \alpha_n)|^{2k} = \sum_{x_1, \ldots, y_k = 1}^{P} e^{2\pi i (\alpha_1 (x_1 + \ldots - y_k) + \ldots + \alpha_n (x_1^n + \ldots - y_k^n))}. \tag{160}$$

Here we unite addends with fixed values of the sums $x_1^\nu + \ldots - y_k^\nu$ $(\nu = 1, 2, \ldots, n)$

$$\left.\begin{array}{c} x_1 + \ldots - y_k = \lambda_1 \\ \ldots\ldots\ldots\ldots\ldots\ldots \\ x_1^n + \ldots - y_k^n = \lambda_n \end{array}\right\}.$$

Since $1 \leqslant x_j \leqslant P$ and $1 \leqslant y_j \leqslant P$ $(j = 1, 2, \ldots, k)$, then by (158) the number of such addends equals $N_{k,n}^{(P)}(\lambda_1, \ldots, \lambda_n)$ and besides

$$|\lambda_\nu| = |x_1^\nu + \ldots + x_k^\nu - y_1^\nu - \ldots - y_k^\nu| < kP^\nu \qquad (\nu = 1, 2, \ldots, n).$$

Therefore,

$$\sum_{x_1, \ldots, y_k = 1}^{P} e^{2\pi i (\alpha_1 (x_1 + \ldots - y_k) + \ldots + \alpha_n (x_1^n + \ldots - y_k^n))}$$
$$= \sum_{\lambda_1, \ldots, \lambda_n} N_{k,n}^{(P)}(\lambda_1, \ldots, \lambda_n) e^{2\pi i (\alpha_1 \lambda_1 + \ldots + \alpha_n \lambda_n)}.$$

The equality (159) follows by (160).

The relation (159) is the expansion of the function $|S(\alpha_1, \ldots, \alpha_n)|^{2k}$ in the multiple Fourier series. The quantities $N_{k,n}^{(P)}(\lambda_1, \ldots, \lambda_n)$ are its Fourier coefficients. Therefore,

$$N_{k,n}^{(P)}(\lambda_1, \ldots, \lambda_n)$$
$$= \int_0^1 \ldots \int_0^1 |S(\alpha_1, \ldots, \alpha_n)|^{2k} e^{-2\pi i (\alpha_1 \lambda_1 + \ldots + \alpha_n \lambda_n)} d\alpha_1 \ldots d\alpha_n \qquad (161)$$

and, in particular,

$$N_{k,n}^{(P)}(0, \ldots, 0) = \int_0^1 \ldots \int_0^1 |S(\alpha_1, \ldots, \alpha_n)|^{2k} d\alpha_1 \ldots d\alpha_n.$$

Hereafter we shall often use the abbreviated notation $N_{k,n}(P)$ and $N_k(P)$ instead of $N_{k,n}^{(P)}(0, \ldots, 0)$, and $N_k^{(P)}(\lambda_1, \ldots, \lambda_n)$ instead of $N_{k,n}^{(P)}(\lambda_1, \ldots, \lambda_n)$. Since the modulus of an integral does not exceed the integral of the modulus of the integrand, then it follows from (161) that under any $\lambda_1, \ldots, \lambda_n$

$$N_k^{(P)}(\lambda_1, \ldots, \lambda_n) \leqslant \int_0^1 \ldots \int_0^1 |S(\alpha_1, \ldots, \alpha_n)|^{2k} d\alpha_1 \ldots d\alpha_n = N_k(P). \qquad (162)$$

Let us show the validity of the following equalities:

$$\sum_{\lambda_n} N_k^{(P)}(\lambda_1,\dots,\lambda_n) = N_k^{(P)}(\lambda_1,\dots,\lambda_{n-1}), \tag{163}$$

$$\sum_{\lambda_1,\dots,\lambda_n} N_k^{(P)}(\lambda_1,\dots,\lambda_n) = P^{2k}, \tag{164}$$

$$\sum_{\lambda_1,\dots,\lambda_n} \left[N_k^{(P)}(\lambda_1,\dots,\lambda_n)\right]^2 = N_{2k}(P). \tag{165}$$

In fact, according to the introduced notation the number of solutions of the system of equations

$$\left.\begin{array}{c} x_1 + \dots - y_k = \lambda_1 \\ \dots\dots\dots\dots\dots\dots \\ x_1^{n-1} + \dots - y_k^{n-1} = \lambda_{n-1} \end{array}\right\} \tag{166}$$

is equal to $N_k^{(P)}(\lambda_1,\dots,\lambda_{n-1})$. Complete this system by the equation

$$x_1^n + \dots - y_k^n = \lambda_n.$$

The number of solutions of the completed system equals $N_k^{(P)}(\lambda_1,\dots,\lambda_n)$. Every solution of the system (166) satisfies one and only one of completed systems arising under distinct values λ_n, and every solution of the completed system satisfies the system (166). Therefore the sum of the quantities $N_k^{(P)}(\lambda_1,\dots,\lambda_n)$ extended to all possible values λ_n is equal to the number of solutions of the system (166), i.e., the equality (163) is valid.

The equality (164) follows immediately from (163):

$$\sum_{\lambda_1,\dots,\lambda_n} N_k^{(P)}(\lambda_1,\dots,\lambda_n) = \sum_{\lambda_1,\dots,\lambda_{n-1}} N_k^{(P)}(\lambda_1,\dots,\lambda_{n-1}) = \dots$$
$$= \sum_{\lambda_1} N_k^{(P)}(\lambda_1) = P^{2k}.$$

To prove the equality (165), we consider the system of equations

$$\left.\begin{array}{c} x_1 + \dots + x_{2k} - y_1 - \dots - y_{2k} = 0 \\ \dots\dots\dots\dots\dots\dots\dots\dots\dots \\ x_1^n + \dots + x_{2k}^n - y_1^n - \dots - y_{2k}^n = 0 \end{array}\right\}. \tag{167}$$

The number of solutions of this system is $N_{2k}(P)$. Collect those solutions, for which under fixed $\lambda_1,\dots,\lambda_n$ the equations

$$\left.\begin{array}{c} x_1 + \dots - y_k = \lambda_1 \\ \dots\dots\dots\dots\dots\dots \\ x_1^n + \dots - y_k^n = \lambda_n \end{array}\right\},$$

$$\left.\begin{array}{l} y_{k+1} + \ldots + y_{2k} - x_{k+1} - \ldots - x_{2k} = \lambda_1 \\ \ldots\ldots\ldots\ldots\ldots\ldots\ldots\ldots\ldots\ldots \\ y_{k+1}^n + \ldots + y_{2k}^n - x_{k+1}^n - \ldots - x_{2k}^n = \lambda_n \end{array}\right\}$$

are fulfilled. Obviously the number of such solutions is equal to $\left[N_k^{(P)}(\lambda_1, \ldots, \lambda_n)\right]^2$. To each n-tuple of values $\lambda_1, \ldots, \lambda_n$ there corresponds one definite aggregate of solutions of the system (167) and each solution of the system enters into one and only one of these aggregates. Thus, considering all possible n-tuples $\lambda_1, \ldots, \lambda_n$, we get all solutions of the system (167) and, therefore,

$$\sum_{\lambda_1, \ldots, \lambda_n} \left[N_k^{(P)}(\lambda_1, \ldots, \lambda_n)\right]^2 = N_{2k}(P).$$

In investigating properties of the system of equations

$$\left.\begin{array}{l} x_1 + \ldots - y_k = \lambda_1 \\ \ldots\ldots\ldots\ldots\ldots \\ x_1^n + \ldots - y_k^n = \lambda_n \end{array}\right\}, \qquad 1 \leqslant x_j, y_j \leqslant P, \tag{168}$$

and in deducing estimates of Weyl's sums, the relationship between the exponential sums

$$S(\alpha_1, \ldots, \alpha_n) = \sum_{x=1}^{P} e^{2\pi i\,(\alpha_1 x + \ldots + \alpha_n x^n)},$$

considered as functions of n variables $\alpha_1, \ldots, \alpha_n$ and the number of solutions of the equation system (168) is used essentially. This relationship is seen from the expansion of the function $|S(\alpha_1, \ldots, \alpha_n)|^{2k}$ in a multiple Fourier series:

$$|S(\alpha_1, \ldots, \alpha_n)|^{2k} = \sum_{\lambda_1, \ldots, \lambda_n = -\infty}^{\infty} N_k^{(P)}(\lambda_1, \ldots, \lambda_n)\, e^{2\pi i\,(\alpha_1 \lambda_1 + \ldots + \alpha_n \lambda_n)}. \tag{169}$$

Actually the series (169), as it was shown in the equality (159), is a finite sum, because if at least one of quantities λ_ν in absolute value is greater than or equal to kP^ν, then the system (168) has no solutions and the corresponding Fourier coefficient vanishes:

$$N_k^{(P)}(\lambda_1, \ldots, \lambda_n) = 0 \qquad (|\lambda_\nu| \geqslant kP^\nu).$$

We shall show that the above established properties (163)–(165) of quantities $N_k^{(P)}(\lambda_1, \ldots, \lambda_n)$ are evident corollaries of this expansion.

In fact, setting $\alpha_1 = \ldots = \alpha_n = 0$ in (169), we obtain the equality (164):

$$P^{2k} = \sum_{\lambda_1, \ldots, \lambda_n} N_k^{(P)}(\lambda_1, \ldots, \lambda_n).$$

The equality (165) follows at once from Parseval's identity for the function $|S(\alpha_1,\dots,\alpha_n)|^{2k}$:

$$\sum_{\lambda_1,\dots,\lambda_n} \left[N_k^{(P)}(\lambda_1,\dots,\lambda_n)\right]^2 = \int_0^1 \dots \int_0^1 \left[|S(\alpha_1,\dots,\alpha_n)|^{2k}\right]^2 d\alpha_1 \dots d\alpha_n$$

$$= \int_0^1 \dots \int_0^1 |S(\alpha_1,\dots,\alpha_n)|^{4k} d\alpha_1 \dots d\alpha_n = N_{2k}(P).$$

Finally, setting $\alpha_n = 0$ in (169), we get

$$|S(\alpha_1,\dots,\alpha_{n-1})|^{2k} = \sum_{\lambda_1,\dots,\lambda_{n-1}} \left[\sum_{\lambda_n} N_k^{(P)}(\lambda_1,\dots,\lambda_n)\right] e^{2\pi i(\alpha_1\lambda_1+\dots+\alpha_{n-1}\lambda_{n-1})}.$$

Hence by virtue of the uniqueness of the expansion of the function $|S(\alpha_1,\dots,\alpha_{n-1})|^{2k}$ in the Fourier series

$$|S(\alpha_1,\dots,\alpha_{n-1})|^{2k} = \sum_{\lambda_1,\dots,\lambda_{n-1}} N_k^{(P)}(\lambda_1,\dots,\lambda_{n-1})\, e^{2\pi i(\alpha_1\lambda_1+\dots+\alpha_{n-1}\lambda_{n-1})}$$

the equality (163)

$$\sum_{\lambda_n} N_k^{(P)}(\lambda_1,\dots,\lambda_n) = N_k^{(P)}(\lambda_1,\dots,\lambda_{n-1})$$

follows.

The most important question in the theory of the systems of equations

$$\left.\begin{array}{c} x_1 + \dots - y_k = 0 \\ \dots\dots\dots\dots\dots\dots \\ x_1^n + \dots - y_k^n = 0 \end{array}\right\}, \qquad 1 \leqslant x_j, y_j \leqslant P,$$

is a question concerning the character of the growth of the number of system solutions in dependence on the magnitude of an interval of the variation of variables, i.e., a question concerning the character of the growth of the quantity $N_k(P)$ while P increases infinitely.

It is easy to establish a lower bound for $N_k(P)$. Indeed, since $1 \leqslant x_j \leqslant P$ ($j = 1,2,\dots,k$), the quantities $x_1,\dots,x_k$ can be chosen in P^k ways. Choosing then $y_1 = x_1, \dots, y_k = x_k$ we obtain P^k solutions. Therefore we have the estimate

$$N_k(P) \geqslant P^k. \tag{170}$$

Next, by (162) and (164)

$$P^{2k} = \sum_{\lambda_1,\ldots,\lambda_n} N_k^{(P)}(\lambda_1,\ldots,\lambda_n) \leqslant N_k(P) \sum_{\substack{\lambda_1,\ldots,\lambda_n \\ |\lambda_\nu|<kP^\nu}} 1 \leqslant (2k)^n P^{\frac{n(n+1)}{2}} N_k(P),$$

and, therefore,

$$N_k(P) \geqslant \frac{1}{(2k)^n} P^{2k-\frac{n(n+1)}{2}}.$$

Taking into account this result and the estimate (170), we get the lower estimate for $N_k(P)$

$$N_k(P) \geqslant \max\left(P^k, \frac{1}{(2k)^n} P^{2k-\frac{n(n+1)}{2}}\right). \tag{171}$$

We shall show that under $k \leqslant n$ this estimate indicates the precise order of the growth of the quantity $N_k(P)$. Indeed, consider the system of equations

$$\left.\begin{array}{c} x_1+\ldots+x_k = y_1+\ldots+y_k \\ \cdots\cdots\cdots\cdots\cdots\cdots \\ x_1^k+\ldots+x_k^k = y_1^k+\ldots+y_k^k \end{array}\right\}, \qquad 1 \leqslant x_j, y_j \leqslant P. \tag{172}$$

In the same way as in the proof of Mordell's lemma (§ 5), it is easy to verify that the quantities $x_1,\ldots,x_k$ satisfying this system coincide with permutations of quantities $y_1,\ldots,y_k$. Hence the number of its solutions does not exceed $k!\,P^k$. Since the system (172) is obtained from the system

$$\left.\begin{array}{c} x_1+\ldots+x_k = y_1+\ldots+y_k \\ \cdots\cdots\cdots\cdots\cdots\cdots \\ x_1^n+\ldots+x_k^n = y_1^n+\ldots+y_k^n \end{array}\right\}, \qquad 1 \leqslant x_j, y_j \leqslant P,$$

by omitting the last $n-k$ equations, then $N_k(P)$ does not exceed the number of solutions of the system (172) and, therefore, $N_k(P) \leqslant k!\,P^k$. Thus, under $k \leqslant n$ the estimate (171) has the precise order with respect to P.

Under $k \geqslant n$ it is easy to show that $N_k(P) \leqslant n!\,P^{2k-n}$. Indeed, using the trivial estimation of the sum $S(\alpha_1,\ldots,\alpha_n)$ we get

$$\begin{aligned} N_k(P) &= \int_0^1 \cdots \int_0^1 |S(\alpha_1,\ldots,\alpha_n)|^{2k} d\alpha_1 \ldots d\alpha_n \\ &\leqslant P^{2k-2n} \int_0^1 \cdots \int_0^1 |S(\alpha_1,\ldots,\alpha_n)|^{2n} d\alpha_1 \ldots d\alpha_n \\ &= P^{2k-2n} N_n(P) \leqslant n!\,P^{2k-n}. \end{aligned} \tag{173}$$

A question on upper estimates of $N_k(P)$, having under $k > n$ a precise order with respect to P, is much more difficult. This question, which is referred to as the mean value theorem, is main in a method suggested by Vinogradov to estimate Weyl's sums.

In proving the mean value theorem we shall need two lemmas.

LEMMA 15. *Under any fixed integer a, the number of solutions of the system of equations*

$$\left.\begin{array}{r}(x_1+a)+\ldots-(y_k+a)=0\\ \ldots\ldots\ldots\ldots\ldots\ldots\ldots\ldots\\ (x_1+a)^n+\ldots-(y_k+a)^n=0\end{array}\right\},\qquad 1\leqslant x_j,y_j\leqslant P, \tag{174}$$

does not depend on a and is equal to $N_k(P)$.

Proof. Let $x_1,\ldots,y_k$ be an arbitrary solution of the system of equations

$$\left.\begin{array}{r}x_1+\ldots-y_k=0\\ \ldots\ldots\ldots\ldots\\ x_1^n+\ldots-y_k^n=0\end{array}\right\},\qquad 1\leqslant x_j,y_j\leqslant P. \tag{175}$$

Then under any $s=1,2,\ldots,n$ we obtain

$$(x_j+a)^s-(y_j+a)^s=\sum_{\nu=0}^{s}C_s^\nu a^{s-\nu}(x_j^\nu-y_j^\nu),$$

$$\sum_{j=1}^{k}\left[(x_j+a)^s-(y_j+a)^s\right]=\sum_{\nu=0}^{s}C_s^\nu a^{s-\nu}\sum_{j=1}^{k}(x_j^\nu-y_j^\nu)=0,$$

where C_s^ν denotes the number of combinations of s objects ν at a time.

Therefore each solution of the system (175) is a solution of the system (174). It is just as easy to verify that in its turn each solution of the system (174) is a solution of the system (175). But then these systems of equations have the same number of solutions, and this is what we had to prove.

Note. According to Lemma 15

$$N_k(P)=\int_0^1\cdots\int_0^1\left|\sum_{x=a+1}^{a+P}e^{2\pi i(\alpha_1x+\ldots+\alpha_nx^n)}\right|^{2k}dx_1\ldots dx_n,$$

and, therefore, under any integer a the equality

$$\int_0^1\cdots\int_0^1\left|\sum_{x=a+1}^{a+P}e^{2\pi i(\alpha_1x+\ldots+\alpha_nx^n)}\right|^{2k}dx_1\ldots dx_n$$
$$=\int_0^1\cdots\int_0^1\left|\sum_{x=1}^{P}e^{2\pi i(\alpha_1x+\ldots+\alpha_nx^n)}\right|^{2k}dx_1\ldots dx_n$$

holds.

Let, as in § 6 (the note of Lemma 7), $T_k(P)$ be the number of solutions of the system of congruences

$$\left.\begin{array}{l} x_1 + \ldots - y_k \equiv 0 \pmod{p} \\ \cdots\cdots\cdots\cdots\cdots\cdots\cdots\cdots \\ x_1^n + \ldots - y_k^n \equiv 0 \pmod{p^n} \end{array}\right\}, \qquad 1 \leqslant x_j, y_j \leqslant P. \tag{176}$$

We shall show that the number of solutions of this system can be expressed in terms of the quantity $N_k^{(P)}(\lambda_1, \ldots, \lambda_n)$.

LEMMA 16. *We have the equality*

$$T_k(P) = \sum_{\lambda_1,\ldots,\lambda_n} N_k^{(P)}(\lambda_1 p, \ldots, \lambda_n p^n),$$

where the summation is extended over the region

$$|\lambda_1| < kPp^{-1}, \ \ldots, \ |\lambda_n| < kPp^{-n}. \tag{177}$$

Proof. It is easily seen that the congruence system (176) is equivalent to the totality of the systems of equations

$$\left.\begin{array}{l} x_1 + \ldots - y_k = \lambda_1 p \\ \cdots\cdots\cdots\cdots\cdots\cdots \\ x_1^n + \ldots - y_k^n = \lambda_n p^n \end{array}\right\}, \qquad 1 \leqslant x_j, y_j \leqslant P,$$

arising under all possible n-tuples of integers $\lambda_1, \ldots, \lambda_n$. Since under fixed values $\lambda_1, \ldots, \lambda_n$ the number of solutions of this system is equal to $N_k^{(P)}(\lambda_1 p, \ldots, \lambda_n p^n)$, then the sum of the quantities $N_k^{(P)}(\lambda_1 p, \ldots, \lambda_n p^n)$, extended over all possible values $\lambda_1, \ldots, \lambda_n$ is equal to the number of solutions of the system of congruences:

$$T_k(P) = \sum_{\lambda_1,\ldots,\lambda_n} N_k^{(P)}(\lambda_1 p, \ldots, \lambda_n p^n).$$

It is sufficient to carry out the summation over the region (177), because otherwise at least for one value ν $(1 \leqslant \nu \leqslant n)$ the inequality $|\lambda_\nu p^\nu| \geqslant kP^\nu$ would be fulfilled and the corresponding summand $N_k^{(P)}(\lambda_1 p, \ldots, \lambda_n p^n)$ would vanish.

§ 13. Vinogradov's mean value theorem

As it was said in the preceding section, the mean value theorem pursues the aim to establish an upper estimate for the quantity $N_k(P)$, where $N_k(P)$ is the number of integral solutions of the system of equations

$$\left.\begin{array}{l} x_1 + \ldots + x_k = y_1 + \ldots + y_k \\ \ldots\ldots\ldots\ldots\ldots\ldots\ldots\ldots \\ x_1^n + \ldots + x_k^n = y_1^n + \ldots + y_k^n \end{array}\right\}, \qquad 1 \leqslant x_j, y_j \leqslant P. \tag{178}$$

The proof of the mean value theorem, suggested by I. M. Vinogradov, is based on a recurrent process reducing the estimation of the quantity $N_k(P)$ to the estimation of $N_{k_1}(P_1)$, where $k_1 < k$ and $P_1 < P$. Two proofs of the mean value theorem are represented below. The first of them is simpler, but it leads to a result, which is valid only under an over-abundant number of variables in the system (178). The second proof is a bit more complicated, but it enables us to obtain results which are close to final ones. Both proofs are carried out with the help of different variants of the p-adic approach suggested by Yu. V. Linnik [34] for this problem.

LEMMA 17. *Let $n \geqslant 2$, $r = n^2$, $P > n^n$, p be a prime, $P^{\frac{1}{n}} \leqslant p < 2P^{\frac{1}{n}}$, and $P_1 = p^{n-1}$. Then under $k > n^2$ for the number of solutions of the system (178) the estimate*

$$N_k(P) \leqslant 2^{2nk} P_1^{2r} p^{2k - \frac{n(n+1)}{2}} N_{k-r}(P_1)$$

holds.

Proof. Let $f(x) = \alpha_1 x + \ldots + \alpha_n x^n$ and the sums S and $S(z)$ be determined with the help of the equalities

$$S = \sum_{x=p+1}^{p+pP_1} e^{2\pi i\, f(x)}, \qquad S(z) = \sum_{\dot{x}=1}^{P_1} e^{2\pi i\, f(z+p\dot{x})}.$$

Then, obviously,

$$S = \sum_{z=1}^{p} \sum_{\dot{x}=1}^{P_1} e^{2\pi i\, f(z+p\dot{x})} = \sum_{z=1}^{p} S(z),$$

$$|S|^{2k} = |S|^{2r}|S|^{2k-2r} \leqslant p^{2k-2r-1}|S|^{2r} \sum_{z=1}^{p} |S(z)|^{2k-2r}.$$

Since the number of solutions of the system (178) grows as P grows, then using the equality (161) and the note of Lemma 15, we obtain

$$N_k(P) \leqslant N_k(pP_1) = \int_0^1 \ldots \int_0^1 |S|^{2k}\, d\alpha_1 \ldots d\alpha_n$$

$$\leqslant p^{2k-2r-1}\sum_{z=1}^{p}\int_0^1\cdots\int_0^1 |S|^{2r}|S(z)|^{2k-2r}d\alpha_1\ldots d\alpha_n. \qquad (179)$$

Let the maximal value of summands in the sum (179) be attained at $z = z_0$. Then we obtain

$$N_k(P)\leqslant p^{2k-2r}\int_0^1\cdots\int_0^1\left|\sum_{x=p+1}^{p+pP_1}e^{2\pi i\,f(x)}\right|^{2r}\left|\sum_{\dot{x}=1}^{P_1}e^{2\pi i\,f(z_0+p\dot{x})}\right|^{2k-2r}d\alpha_1\ldots d\alpha_n.$$

It is easily seen that the integral in this estimate is equal to the number of solutions of the system of equations

$$\left.\begin{array}{c} x_1+\ldots-y_r=(z_0+p\dot{x}_1)+\ldots-(z_0+p\dot{y}_{k-r})\\ \cdots\cdots\cdots\cdots\cdots\cdots\cdots\cdots\cdots\cdots\\ x_1^n+\ldots-y_r^n=(z_0+p\dot{x}_1)^n+\ldots-(z_0+p\dot{y}_{k-r})^n\end{array}\right\},$$

$$p<x_j,y_j\leqslant p+pP_1,\quad 1\leqslant\dot{x}_j,\dot{y}_j\leqslant P_1,$$

or, this is just the same, to the number of solutions of the system

$$\left.\begin{array}{c} (z_0+x_1)+\ldots-(z_0+y_r)=(z_0+p\dot{x}_1)+\ldots-(z_0+p\dot{y}_{k-r})\\ \cdots\cdots\cdots\cdots\cdots\cdots\cdots\cdots\cdots\cdots\cdots\cdots\\ (z_0+x_1)^n+\ldots-(z_0+y_r)^n=(z_0+p\dot{x}_1)^n+\ldots-(z_0+p\dot{y}_{k-r})^n\end{array}\right\},$$

$$p-z_0<x_j,y_j\leqslant p-z_0+pP_1,\quad 1\leqslant\dot{x}_j,\dot{y}_j\leqslant P_1.$$

In its turn this system is equivalent (see Lemma 15) to the system

$$\left.\begin{array}{c} x_1+\ldots-y_r=p(\dot{x}_1+\ldots-\dot{y}_{k-r})\\ \cdots\cdots\cdots\cdots\cdots\cdots\\ x_1^n+\ldots-y_r^n=p^n(\dot{x}_1^n+\ldots-\dot{y}_{k-r}^n)\end{array}\right\},$$

$$p-z_0<x_j,y_j\leqslant p-z_0+pP_1,\quad 1\leqslant\dot{x}_j,\dot{y}_j\leqslant P_1.$$

Let us replace the interval of variation of x_j and y_j by a wider one:

$$1\leqslant x_j,y_j\leqslant 2pP_1\qquad (j=1,2,\ldots,r).$$

Then, collecting solutions with fixed values of the sums $\dot{x}_1^\nu+\ldots-\dot{y}_{k-r}^\nu$ $(\nu=1,2,\ldots,n)$ and using the estimate (162), we get

$$N_k(P)\leqslant p^{2k-2r}\sum_{\lambda_1,\ldots,\lambda_n}N_{k-r}^{(P_1)}(\lambda_1,\ldots,\lambda_n)N_r^{(2pP_1)}(\lambda_1p,\ldots,\lambda_np^n)$$

$$\leqslant p^{2k-2r}N_{k-r}(P_1)\sum_{\lambda_1,\ldots,\lambda_n}N_r^{(2pP_1)}(\lambda_1p,\ldots,\lambda_np^n),$$

where the summation is extended over the region $|\lambda_\nu| < r(2P_1)^\nu$ $(\nu = 1, 2, \ldots, n)$. Hence, since $P_1 = p^{n-1}$, using Lemma 16 and the note (90), we get the lemma assertion:

$$\begin{aligned} N_k(P) &\leqslant p^{2k-2r} N_{k-r}(P_1)\, T_r(2p^n) \\ &\leqslant (2n)^{2r} p^{2k+2(n-1)r-\frac{n(n+1)}{2}} N_{k-r}(P_1) \\ &\leqslant 2^{2nk} P_1^{2r} p^{2k-\frac{n(n+1)}{2}} N_{k-r}(P_1). \end{aligned}$$

THEOREM 15. *Let* $n \geqslant 2,\ \tau \geqslant 0,\ k > n^2\tau$,

$$P > n^{n\left(1+\frac{1}{n-1}\right)^\tau}, \qquad \varepsilon_\tau = \frac{n(n+1)}{2}\left(1 - \frac{1}{n}\right)^\tau.$$

Then for the number of solutions of the system (178) *we have the estimate*

$$N_k(P) \leqslant 2^{4nk\tau} P^{2k-\frac{n(n+1)}{2}+\varepsilon_\tau}. \tag{180}$$

Proof. The assertion of the theorem can be obtained with the help of rather simple induction. Indeed, under $\tau = 0$ the estimate (180) is trivial. Let it be true under a certain $\tau \geqslant 0$. Choose

$$k > n^2(\tau + 1) \quad \text{and} \quad P > n^{n\left(1+\frac{1}{n-1}\right)^{\tau+1}}.$$

Determine positive integers r, P_1, and a prime p as in Lemma 17:

$$r = n^2, \qquad P^{\frac{1}{n}} \leqslant p < 2P^{\frac{1}{n}}, \qquad P_1 = p^{n-1}.$$

Then

$$k - r > n^2\tau, \qquad P_1 > P^{\frac{n-1}{n}} > n^{n\left(1+\frac{1}{n-1}\right)^\tau}$$

and by the induction hypothesis

$$N_{k-r}(P_1) \leqslant 2^{4n(k-r)\tau} P_1^{2k-2r-\frac{n(n+1)}{2}+\varepsilon_\tau}. \tag{181}$$

But according to Lemma 17

$$N_k(P) \leqslant 2^{2nk} P_1^{2r} p^{2k-\frac{n(n+1)}{2}} N_{k-r}(P_1)$$

and, therefore, by (181)

$$N_k(P) \leqslant 2^{4nk\tau+2nk}(pP_1)^{2k-\frac{n(n+1)}{2}} P_1^{\varepsilon_\tau}.$$

Hence, since $pP_1 = p^n < 2^n P$, $(n-1)\varepsilon_\tau = n\varepsilon_{\tau+1}$, and

$$P_1^{\varepsilon_\tau} = p^{(n-1)\varepsilon_\tau} \leqslant (2^n P)^{\varepsilon_{\tau+1}},$$

we get

$$N_k(P) \leqslant 2^{4nk\tau+2nk}(2^n P)^{2k-\frac{n(n+1)}{2}+\varepsilon_{\tau+1}} \leqslant 2^{4nk(\tau+1)} P^{2k-\frac{n(n+1)}{2}+\varepsilon_{\tau+1}}.$$

The theorem is proved.

Now we shall consider the question about the precision of the obtained results. Since

$$\left(1-\frac{1}{n}\right)^\tau \leqslant e^{-\frac{\tau}{n}},$$

then under the growth of τ and the corresponding growth of k, the quantity ε_τ tends to zero. So under $\tau > 2n\log(n+1)$ we get

$$\varepsilon_\tau = \frac{n(n+1)}{2}\left(1-\frac{1}{n}\right)^\tau < \frac{1}{2}.$$

Respectively, under $\tau > 3n\log(n+1)$ we have $\varepsilon_\tau < \frac{1}{2(n+1)}$. Hence it follows that for any $\varepsilon > 0$ under $k > 3n^3\log(n+1)$ and $n \geqslant \frac{1}{2\varepsilon}$ we have the estimate

$$N_k(P) \leqslant 2^{\frac{2}{n}k^2} P^{2k-\frac{n(n+1)}{2}+\varepsilon} = O\left(P^{2k-\frac{n(n+1)}{2}+\varepsilon}\right). \tag{182}$$

On the other hand, by (171)

$$N_k(P) \geqslant \frac{1}{(2k)^n} P^{2k-\frac{n(n+1)}{2}}.$$

It is seen from comparison of this estimate and the estimate (182), that the order of the estimate (180) is almost best possible.

A question about the least value of k, under which the estimate (180) is fulfilled, is much more difficult. This question is important in connection with the following circumstance: estimates of Weyl's sums obtained by the help of the mean value theorem are, as a rule, more precise, if one succeeds to establish an estimate of the form (180) under lesser values of k, i.e., the lesser the better.

Let us show that the estimate

$$N_k(P) = O\left(P^{2k-\frac{n(n+1)}{2}+\varepsilon}\right) \tag{183}$$

cannot be fulfilled under $k < \frac{n(n+1)}{2}$. Indeed, according to (171) $N_k(P) \geqslant P^k$ and, therefore, in order to satisfy the estimate (183), the estimate

$$P^k = O\left(P^{2k-\frac{n(n+1)}{2}+\varepsilon}\right)$$

should be fulfilled, but that is possible only under $k \geqslant \frac{n(n+1)}{2}$. Thus the best result which might be expected to obtain is getting a precise estimate (with respect to the order) under $k = \frac{n(n+1)}{2}$. The estimate (183) following from Theorem 15 was obtained under $k \geqslant 3n^3 \log(n+1)$. Using the Linnik lemma (Lemma 9) instead of Lemma 7, we get now this estimate under $k \geqslant 3n^2 \log(n+1)$.

LEMMA 18. *Let $n \geqslant 2$, $P \geqslant (2n)^{2n}$, p be a prime, $\frac{1}{2} P^{\frac{1}{n}} \leqslant p < P^{\frac{1}{n}}$, and $P_1 = [Pp^{-1}] + 1$. Then under $k \geqslant \frac{n(n+1)}{2}$ we have the estimate*

$$N_k(P) \leqslant 2(2k)^{2n} P_1^{2n} p^{2k-\frac{n(n+1)}{2}} N_{k-n}(P_1). \tag{184}$$

Proof. As in Lemma 17, we introduce the notation $f(x) = \alpha_1 x + \ldots + \alpha_n x^n$,

$$S = \sum_{x=p+1}^{p+pP_1} e^{2\pi i f(x)}, \qquad S(z) = \sum_{\dot{x}=1}^{P_1} e^{2\pi i f(z+p\dot{x})}.$$

Then we get

$$|S|^{2k} = \left| \sum_{z_1,\ldots,z_k=1}^{P} S(z_1)\ldots S(z_k) \right|^2.$$

We shall say that a k-tuple $z_1, \ldots, z_k$ belongs to the first class, if it is possible to find n distinct quantities z_j in it. All remaining k-tuples are to be of the second class. Since

$$\begin{aligned} |S|^{2k} &= \left| \sum\nolimits_{1\, z_1,\ldots,z_k} S(z_1)\ldots S(z_k) + \sum\nolimits_{2\, z_1,\ldots,z_k} S(z_1)\ldots S(z_k) \right|^2 \\ &\leqslant 2\left| \sum\nolimits_{1\, z_1,\ldots,z_k} S(z_1)\ldots S(z_k) \right|^2 + 2\left| \sum\nolimits_{2\, z_1,\ldots,z_k} S(z_1)\ldots S(z_k) \right|^2, \end{aligned}$$

where the sums $\sum_1$ and $\sum_2$ are over k-tuples of the first and the second classes, respectively, then observing that $P \leqslant pP_1$ and using the note of Lemma 15, we obtain

$$N_k(P) \leqslant N_k(pP_1) = \int_0^1 \cdots \int_0^1 |S|^{2k} d\alpha_1 \ldots d\alpha_n$$

$$\leqslant 2\int_0^1 \cdots \int_0^1 \left| \sum_{z_1,\ldots,z_k}{}_1 S(z_1)\ldots S(z_k) \right|^2 d\alpha_1 \ldots d\alpha_n$$

$$+ 2\int_0^1 \cdots \int_0^1 \left| \sum_{z_1,\ldots,z_k}{}_2 S(z_1)\ldots S(z_k) \right|^2 d\alpha_1 \ldots d\alpha_n. \tag{185}$$

Denote by N_k' and N_k'' the integrals of the right-hand side of this inequality. Hereafter we shall designate the number of combinations of k objects n at a time by C_k^n. Since n distinct quantities can be arranged on k places in C_k^n ways, then

$$N_k' \leqslant (C_k^n)^2 \int_0^1 \cdots \int_0^1 \left| \sum_{z_1,\ldots,z_k}{}_1^* S(z_1)\ldots S(z_k) \right|^2 d\alpha_1 \ldots d\alpha_n,$$

where in the sum $\sum_1^*$ distinct z_j occupy the first n places and the variables $z_{n+1},\ldots,z_k$ independently run over the interval $[1,p]$. Hence, observing that

$$\left| \sum_{z_1,\ldots,z_k}{}_1 S(z_1)\ldots S(z_k) \right|^2 = \left| \sum_{z_1,\ldots,z_n}{}_1^* S(z_1)\ldots S(z_n) \right|^2 \left| \sum_{z=1}^{p} S(z) \right|^{2(k-n)}$$

$$\leqslant p^{2(k-n)-1} \left| \sum_{z_1,\ldots,z_n}{}_1^* S(z_1)\ldots S(z_n) \right|^2 \sum_{z=1}^{p} |S(z)|^{2(k-n)},$$

we get

$$N_k' \leqslant (C_k^n)^2 p^{2(k-n)-1} \sum_{z=1}^{p} \int_0^1 \cdots \int_0^1 \left| \sum_{z_1,\ldots,z_n}{}_1 S(z_1)\ldots S(z_n) \right|^2 |S(z)|^{2(k-n)} d\alpha_1 \ldots d\alpha_n.$$

It is easily seen that under a fixed z the integral in this estimate is equal to the number of solutions of the system of equations

$$\left.\begin{array}{c} (z_1 + px_1') + \ldots - (t_n + py_n') = (z + p\dot{x}_1) + \ldots - (z + p\dot{y}_{k-n}) \\ \cdots\cdots\cdots\cdots\cdots\cdots\cdots\cdots\cdots\cdots \\ (z_1 + px_1')^n + \ldots - (t_n + py_n')^n = (z + p\dot{x}_1)^n + \ldots - (z + p\dot{y}_{k-n})^n \end{array}\right\}, \tag{186}$$

where $1 \leqslant x'_j, y'_j, \dot{x}_j, \dot{y}_j \leqslant P_1$, $1 \leqslant z_j, t_j \leqslant p$ and under $i \neq j$ the conditions $z_i \neq z_j$, $t_i \neq t_j$ are fulfilled. Introduce new variables x_j and y_j

$$z_j + px'_j = z + x_j, \qquad t_j + py'_j = z + y_j \qquad (j = 1, 2, \ldots, n).$$

Then the system (186) takes on the form

$$\left.\begin{array}{c} (z+x_1) + \ldots - (z+y_n) = (z+p\dot{x}_1) + \ldots - (z+p\dot{y}_{k-n}) \\ \cdots\cdots\cdots\cdots\cdots\cdots\cdots\cdots\cdots\cdots \\ (z+x_1)^n + \ldots - (z+y_n)^n = (z+p\dot{x}_1)^n + \ldots - (z+p\dot{y}_{k-n})^n \end{array}\right\}, \tag{187}$$

and the region of variables variation is determined by the conditions $1 \leqslant \dot{x}_j, \dot{y}_j \leqslant P_1$, $p - z < x_j, y_j \leqslant p - z + pP_1$ and $x_i \not\equiv x_j$, $y_i \not\equiv y_j \pmod p$ under $i \neq j$.

By (174) the number of solutions of the system (187) does not exceed the number of solutions of the system of equations

$$\left.\begin{array}{c} x_1 + \ldots - y_n = p(\dot{x}_1 + \ldots - \dot{y}_{k-n}) \\ \cdots\cdots\cdots\cdots\cdots\cdots \\ x_1^n + \ldots - y_n^n = p^n(\dot{x}_1^n + \ldots - \dot{y}_{k-n}^n) \end{array}\right\}, \qquad \begin{array}{l} 1 \leqslant x_j, y_j \leqslant pP_1 + p, \\ i \neq j \Rightarrow x_i \not\equiv x_j,\ y_i \not\equiv y_j \pmod p, \\ 1 \leqslant \dot{x}_j, \dot{y}_j \leqslant P_1. \end{array}$$

Collecting solutions with fixed values of sums $\dot{x}_1^\nu + \ldots - \dot{y}_{k-n}^\nu$ $(\nu = 1, 2, \ldots, n)$ we obtain

$$\begin{aligned} N'_k &\leqslant \left(C_k^n\right)^2 p^{2(k-n)} \sum_{\lambda_1, \ldots, \lambda_n} N_{k-n}^{(P_1)}(\lambda_1, \ldots, \lambda_n) N_n^*(\lambda_1 p, \ldots, \lambda_n p^n) \\ &\leqslant \left(C_k^n\right)^2 p^{2(k-n)} N_{k-n}(P_1) \sum_{\lambda_1, \ldots, \lambda_n} N_n^*(\lambda_1 p, \ldots, \lambda_n p^n), \end{aligned}$$

where $N_n^*(\lambda_1 p, \ldots, \lambda_n p^n)$ is the number of solutions of the system

$$\left.\begin{array}{c} x_1 + \ldots - y_n = \lambda_1 p \\ \cdots\cdots\cdots\cdots \\ x_1^n + \ldots - y_n^n = \lambda_n p^n \end{array}\right\}, \qquad \begin{array}{l} 1 \leqslant x_i, y_i \leqslant pP_1 + p, \\ i \neq j \Rightarrow x_i \not\equiv x_j,\ y_i \not\equiv y_j \pmod p, \end{array}$$

and the summation is over the region $|\lambda_\nu| < n(P_1 + 1)^\nu$ $(\nu = 1, 2, \ldots, n)$. But according to Lemma 16

$$\sum_{\lambda_1, \ldots, \lambda_n} N_n^*(\lambda_1 p, \ldots, \lambda_n p^n) = T_n^*(pP_1 + p) \leqslant T_n^*(mp^n),$$

where

$$m = \left[\frac{pP_1 + p}{p^n}\right] + 1 \leqslant 2\, P_1 p^{-(n-1)}$$

and $T_n^*(mp^n)$ is the number of solutions of the system of congruences

$$\left.\begin{array}{l} x_1 + \ldots - y_n \equiv 0 \pmod{p} \\ \cdots\cdots\cdots\cdots\cdots\cdots\cdots\cdots \\ x_1^n + \ldots - y_n^n \equiv 0 \pmod{p^n} \end{array}\right\}, \qquad \begin{array}{l} 1 \leqslant x_j, y_j \leqslant mp^n, \\ i \neq j \Rightarrow x_i \not\equiv x_j,\ y_i \not\equiv y_j \pmod{p}. \end{array}$$

Therefore, using the estimate (93), we obtain

$$\begin{aligned} T_n^*(mp^n) &\leqslant n!\, m^{2n} p^{2n^2 - \frac{n(n+1)}{2}} \leqslant n!\, 2^{2n} P_1^{2n} p^{2n - \frac{n(n+1)}{2}}, \\ N_k' &\leqslant \left(C_k^n\right)^2 p^{2(k-n)} N_{k-n}(P_1)\, T_n^*(mp^n) \\ &\leqslant \frac{1}{2}(2k)^{2n} P_1^{2n} p^{2k - \frac{n(n+1)}{2}} N_{k-n}(P_1). \end{aligned} \tag{188}$$

Now we shall estimate the quantity N_k''. Observing that the number of k-tuples of the second class does not exceed $n^k p^{n-1}$, we get

$$\left| \sum_{z_1,\ldots,z_k} {}_2\, S(z_1)\ldots S(z_k) \right|^2 \leqslant n^{2k} p^{2n-2} \sum_{z=1}^{p} |S(z)|^{2k} \leqslant n^{2k} p^{2n-2} P_1^{2n} \sum_{z=1}^{p} |S(z)|^{2k-2n},$$

$$N_k'' \leqslant n^{2k} P_1^{2n} p^{2n-2} \sum_{z=1}^{p} \int_0^1 \ldots \int_0^1 |S(z)|^{2k-2n} d\alpha_1 \ldots d\alpha_n = n^{2k} P_1^{2n} p^{2n-1} N_{k-n}(P_1).$$

Since by the lemma conditions $k \geqslant \frac{n(n+1)}{2}$ and $p > n^2$, then

$$n^{2k} p^{2n-1} \leqslant n^{4n-2} p^{2k - \frac{n(n+1)}{2}} \leqslant \frac{1}{2}(2k)^{2n} p^{2k - \frac{n(n+1)}{2}},$$

and, therefore,

$$N_k'' \leqslant \frac{1}{2}(2k)^{2n} P_1^{2n} p^{2k - \frac{n(n+1)}{2}} N_{k-n}(P_1). \tag{189}$$

Now we obtain the lemma assertion from (185), (188), and (189)

$$N_k(P) \leqslant 2N_k' + 2N_k'' \leqslant 2(2k)^{2n} P_1^{2n} p^{2k - \frac{n(n+1)}{2}} N_{k-n}(P_1).$$

The recurrent inequality (184) enables us to make the statement of the mean value theorem essentially stronger, because this inequality reduces the estimation of $N_k(P)$ to the estimation for $N_{k-n}(P_1)$ (but not to $N_{k-n^2}(P_1)$ as it was obtained earlier in Lemma 17).

THEOREM 16. *Let $n \geqslant 2$, $\tau \geqslant 0$, $k = \frac{n(n+1)}{2} + n\tau$, and*

$$\varepsilon_\tau = \frac{n(n-1)}{2}\left(1 - \frac{1}{n}\right)^\tau.$$

Then for the number of solutions of the system (178) the estimate

$$N_k(P) \leqslant (2k)^{2k}(2n)^{n^3} P^{2k - \frac{n(n+1)}{2} + \varepsilon_\tau} \tag{190}$$

holds under any $P \geqslant 1$.

Proof. Since, obviously,

$$\begin{aligned} \varepsilon_0 + \ldots + \varepsilon_{\tau-1} &= \frac{n(n-1)}{2} \sum_{\nu=0}^{\tau-1} \left(1 - \frac{1}{n}\right)^\nu \\ &= \frac{n^2(n-1)}{2}\left[1 - \left(1 - \frac{1}{n}\right)^\tau\right] < \frac{1}{2} n^3, \end{aligned}$$

then to prove the estimate (190) it suffices to show that

$$N_k(P) \leqslant (2k)^{2k}(2n)^{2(\varepsilon_0 + \ldots + \varepsilon_{\tau-1})} P^{2k - \frac{n(n+1)}{2} + \varepsilon_\tau}. \tag{191}$$

If $\tau = 0$, then this estimate takes on the form

$$N_k(P) \leqslant (2k)^{2k} P^{2k-n}$$

and is fulfilled by (173) under any $P \geqslant 1$. Apply the induction. Let under a certain $\tau \geqslant 0$ and $k = \frac{n(n+1)}{2} + n\tau$ the estimate (191) be fulfilled under any $P \geqslant 1$. Prove it for $\tau + 1$, i.e., under $k = \frac{n(n+1)}{2} + n(\tau + 1)$. We shall consider the cases $P \geqslant (2n)^{2n}k^2$ and $P < (2n)^{2n}k^2$ separately.

If $P \geqslant (2n)^{2n}k^2$, then by Lemma 18

$$N_k(P) \leqslant 2\,(2k)^{2n} P_1^{2n} p^{2k - \frac{n(n+1)}{2}} N_{k-n}(P_1),$$

where $\frac{1}{2} P^{\frac{1}{n}} \leqslant p < P^{\frac{1}{n}}$ and $P_1 = \left[Pp^{-1}\right] + 1$. Since $k - n = \frac{n(n+1)}{2} + n\tau$, then using the induction hypothesis, we obtain

$$N_{k-n}(P_1) \leqslant (2k - 2n)^{2k-2n}(2n)^{2(\varepsilon_0 + \ldots + \varepsilon_{\tau-1})} P_1^{2k - 2n - \frac{n(n+1)}{2} + \varepsilon_\tau},$$

$$N_k(P) \leqslant 2\,(2k)^{2k}\left(1 - \frac{n}{k}\right)^{2k-2n}(2n)^{2(\varepsilon_0 + \ldots + \varepsilon_{\tau-1})}(pP_1)^{2k - \frac{n(n+1)}{2}} P_1^{\varepsilon_\tau}. \tag{192}$$

Observing that $P > 4\,k^2$ and, therefore,

$$pP_1 < P\left(1 + \frac{1}{P^{1-\frac{1}{n}}}\right) < P\left(1 + \frac{1}{2k}\right),$$

$$P_1 < 2\,P^{1-\frac{1}{n}} + 1 < 2\,P^{1-\frac{1}{n}}\left(1 + \frac{1}{2k}\right),$$

we get

$$(pP_1)^{2k - \frac{n(n+1)}{2}} P_1^{\varepsilon_\tau} < 2^{\varepsilon_\tau} P^{2k - \frac{n(n+1)}{2} + \left(1 - \frac{1}{n}\right)\varepsilon_\tau} \left(1 + \frac{1}{2k}\right)^{2k-n}$$
$$< 3 \cdot 2^{\varepsilon_\tau} P^{2k - \frac{n(n+1)}{2} + \varepsilon_{\tau+1}}.$$

But then it follows from (192) that

$$N_k(P) < 6\,e^{-\frac{2n(k-n)}{k}} 2^{\varepsilon_\tau} (2n)^{2(\varepsilon_0 + \ldots + \varepsilon_{\tau-1})} (2k)^{2k} P^{2k - \frac{n(n+1)}{2} + \varepsilon_{\tau+1}}$$
$$< (2k)^{2k} (2n)^{2(\varepsilon_0 + \ldots + \varepsilon_\tau)} P^{2k - \frac{n(n+1)}{2} + \varepsilon_{\tau+1}}. \qquad (193)$$

Let now $P < (2n)^{2n} k^2$. By the induction hypothesis

$$N_{k-n}(P) \leqslant (2k - 2n)^{2k-2n} (2\,n)^{2(\varepsilon_0 + \ldots + \varepsilon_{\tau-1})} P^{2k - 2n - \frac{n(n+1)}{2} + \varepsilon_\tau}$$

and, therefore,

$$N_k(P) \leqslant P^{2n} N_{k-n}(P)$$
$$\leqslant (2k)^{2k-2n} (2n)^{2(\varepsilon_0 + \ldots + \varepsilon_{\tau-1})} P^{\varepsilon_\tau - \varepsilon_{\tau+1}} P^{2k - \frac{n(n+1)}{2} + \varepsilon_{\tau+1}}.$$

Hence, observing that

$$P^{\varepsilon_\tau - \varepsilon_{\tau+1}} < \left[(2n)^{2n} k^2\right]^{\frac{1}{n}\varepsilon_\tau} \leqslant k^{n-1} (2n)^{2\varepsilon_\tau},$$

for values P less than $k^2(2n)^{2n}$, we get the estimate (191) too. Thus the estimate is fulfilled for any $P \geqslant 1$, and the theorem is proved completely.

Let now $k \geqslant k_0$, where $k_0 = \frac{n(n+1)}{2} + n\tau$. It is easy to verify that the estimate (190) proved in Theorem 16 for $k = k_0$ holds under $k > k_0$ as well. It suffices to use the evident inequality

$$N_k(P) \leqslant P^{2k - 2k_0} N_{k_0}(P)$$

and apply the estimate (190) to $N_{k_0}(P)$.

Note. With the help of more complicated considerations [44] the mean value theorem can be improved by removing the factor P^{ε_τ} and so it is possible to get under $n \geqslant 1,\ k > cn^2 \log n,\ P \geqslant 1$ the estimate

$$N_k(P) \leqslant C(n) P^{2k - \frac{n(n+1)}{2}}, \tag{194}$$

where c is an absolute constant and $C(n)$ is a constant depending on n only. An elementary proof of the mean value theorem in the form (190) is obtained in the article [37].

§ 14. Estimates of Weyl's sums

To obtain estimates of Weyl's sums by the Vinogradov method besides the mean value theorem we need two comparatively simple lemmas.

LEMMA 19. *Let $f(x)$ be an arbitrary function taking on real values. Then under any positive integers P, P_1, a, and k we have:*

$$1^\circ. \quad \left|\sum_{x=1}^{P} e^{2\pi i\, f(x)}\right| \leqslant \frac{1}{P_1} \sum_{y=0}^{P_1-1} \left|\sum_{x=1}^{P} e^{2\pi i\, f(x+y)}\right| + P_1 - 1,$$

$$2^\circ. \quad \left|\sum_{x=1}^{P} e^{2\pi i\, f(x)}\right| \leqslant \frac{1}{P_1^2} \sum_{x=1}^{P} \left|\sum_{y,z=1}^{P_1} e^{2\pi i\, f(x+ayz)}\right| + 2\,aP_1^2,$$

$$3^\circ. \quad \left|\sum_{x=1}^{P} e^{2\pi i\, f(x)}\right|^{2k+1} \leqslant 2^{2k+1} \sum_{y=0}^{P-1} \left|\sum_{x=1}^{P} e^{2\pi i\, f(x+y)}\right|^{2k}. \tag{195}$$

Proof. Under any integer $y \geqslant 0$

$$\begin{aligned} \sum_{x=1}^{P} e^{2\pi i\, f(x)} &= \sum_{x=1}^{y} e^{2\pi i\, f(x)} + \sum_{x=y+1}^{y+P} e^{2\pi i\, f(x)} - \sum_{x=P+1}^{P+y} e^{2\pi i\, f(x)} \\ &= \sum_{x=1}^{P} e^{2\pi i\, f(x+y)} + 2\theta_y y, \end{aligned} \tag{196}$$

where $|\theta_y| \leqslant 1$. Hence, carrying out the summation over y, we get the assertion 1°:

$$\left|\sum_{x=1}^{P} e^{2\pi i\, f(x)}\right| \leqslant \left|\sum_{x=1}^{P} e^{2\pi i\, f(x+y)}\right| + 2y,$$

$$P_1 \left|\sum_{x=1}^{P} e^{2\pi i\, f(x)}\right| \leqslant \sum_{y=0}^{P_1-1} \left|\sum_{x=1}^{P} e^{2\pi i\, f(x+y)}\right| + P_1(P_1 - 1).$$

To prove the estimate 2° we replace y by ayz in the equality (196) and carry out summing with respect to y and z:

$$\sum_{x=1}^{P} e^{2\pi i\, f(x)} = \sum_{x=1}^{P} e^{2\pi i\, f(x+ayz)} + 2\theta(y,z)\, ayz,$$

$$P_1^2 \sum_{x=1}^{P} e^{2\pi i\, f(x)} = \sum_{x=1}^{P} \sum_{y,z=1}^{P_1} e^{2\pi i\, f(x+ayz)} + 2a \sum_{y,z=1}^{P_1} \theta(y,z)\, yz.$$

Hence, because $|\theta(y,z)| \leqslant 1$ and

$$\sum_{y,z=1}^{P_1} yz = \frac{P_1^2(P_1+1)^2}{4} \leqslant P_1^4,$$

the assertion 2° follows:

$$P_1^2 \left| \sum_{x=1}^{P} e^{2\pi i\, f(x)} \right| \leqslant \sum_{x=1}^{P} \left| \sum_{y,z=1}^{P_1} e^{2\pi i\, f(x+ayz)} \right| + 2aP_1^4.$$

Determine S_1 and P_1 with the help of the equalities

$$S_1 = \sum_{y=0}^{P-1} \left| \sum_{x=1}^{P} e^{2\pi i\, f(x+y)} \right|^{2k}, \qquad P_1 = \min\left(\left[S_1^{\frac{1}{2k+1}} \right] + 1,\, P \right).$$

Then $S_1^{\frac{1}{2k+1}} \leqslant P_1 \leqslant P$ and, therefore,

$$\left(\frac{1}{P_1} \sum_{y=0}^{P_1-1} \left| \sum_{x=1}^{P} e^{2\pi i\, f(x+y)} \right| \right)^{2k} \leqslant \frac{1}{P_1} \sum_{y=0}^{P_1-1} \left| \sum_{x=1}^{P} e^{2\pi i\, f(x+y)} \right|^{2k} \leqslant \frac{1}{P_1} S_1,$$

$$\frac{1}{P_1} \sum_{y=0}^{P_1-1} \left| \sum_{x=1}^{P} e^{2\pi i\, f(x+y)} \right| \leqslant \left(\frac{1}{P_1} S_1 \right)^{\frac{1}{2k}} \leqslant S_1^{\frac{1}{2k+1}}.$$

Hence, using the estimate 1°, we get the assertion 3°:

$$\left| \sum_{x=1}^{P} e^{2\pi i\, f(x)} \right| \leqslant \frac{1}{P_1} \sum_{y=0}^{P_1-1} \left| \sum_{x=1}^{P} e^{2\pi i\, f(x+y)} \right| + P_1 - 1 \leqslant 2\, S_1^{\frac{1}{2k+1}},$$

$$\left| \sum_{x=1}^{P} e^{2\pi i\, f(x)} \right|^{2k+1} \leqslant 2^{2k+1} S_1 = 2^{2k+1} \sum_{y=0}^{P-1} \left| \sum_{x=1}^{P} e^{2\pi i\, f(x+y)} \right|^{2k}.$$

LEMMA 20. *If a function $F(\alpha_1, \ldots, \alpha_n)$ is given by the multiple Fourier expansion*

$$F(\alpha_1, \ldots, \alpha_n) = \sum_{\lambda_1, \ldots, \lambda_n = -\infty}^{\infty} C(\lambda_1, \ldots, \lambda_n) e^{2\pi i (\alpha_1 \lambda_1 + \ldots + \alpha_n \lambda_n)}$$

and satisfying the condition

$$F(\alpha_1, \ldots, \alpha_n) \geqslant 0,$$

then under any positive integers $q_1, \ldots, q_n$ we have

$$F(\alpha_1, \ldots, \alpha_n) \leqslant q_1 \ldots q_n \sum_{\lambda_1, \ldots, \lambda_n = -\infty}^{\infty} C(\lambda_1 q_1, \ldots, \lambda_n q_n) e^{2\pi i (\alpha_1 q_1 \lambda_1 + \ldots + \alpha_n q_n \lambda_n)}.$$

Proof. Since $F(\alpha_1, \ldots, \alpha_n) \geqslant 0$, then

$$F(\alpha_1, \ldots, \alpha_n) \leqslant \sum_{x_1=0}^{q_1-1} \cdots \sum_{x_n=0}^{q_n-1} F\left(\alpha_1 + \frac{x_1}{q_1}, \ldots, \alpha_n + \frac{x_n}{q_n}\right) \tag{197}$$

$$= \sum_{\lambda_1, \ldots, \lambda_n = -\infty}^{\infty} C(\lambda_1, \ldots, \lambda_n) e^{2\pi i (\alpha_1 \lambda_1 + \ldots + \alpha_n \lambda_n)} \sum_{x_1=0}^{q_1-1} \cdots \sum_{x_n=0}^{q_n-1} e^{2\pi i \left(\frac{\lambda_1 x_1}{q_1} + \ldots + \frac{\lambda_n x_n}{q_n}\right)}.$$

By Lemma 2

$$\sum_{x_1=0}^{q_1-1} \cdots \sum_{x_n=0}^{q_n-1} e^{2\pi i \left(\frac{\lambda_1 x_1}{q_1} + \ldots + \frac{\lambda_n x_n}{q_n}\right)} = q_1 \ldots q_n \delta_{q_1}(\lambda_1) \ldots \delta_{q_n}(\lambda_n).$$

Using this equality, we obtain the lemma assertion from (197):

$$\begin{aligned} &F(\alpha_1, \ldots, \alpha_n) \\ &\leqslant q_1 \ldots q_n \sum_{\lambda_1, \ldots, \lambda_n = -\infty}^{\infty} C(\lambda_1, \ldots, \lambda_n) e^{2\pi i (\alpha_1 \lambda_1 + \ldots + \alpha_n \lambda_n)} \delta_{q_1}(\lambda_1) \ldots \delta_{q_n}(\lambda_n) \\ &= q_1 \ldots q_n \sum_{\lambda_1, \ldots, \lambda_n = -\infty}^{\infty} C(q_1 \lambda_1, \ldots, q_n \lambda_n) e^{2\pi i (\alpha_1 q_1 \lambda_1 + \ldots + \alpha_n q_n \lambda_n)}. \end{aligned}$$

COROLLARY. *Let $f(x) = \alpha_1 x + \ldots + \alpha_n x^n$ and*

$$S(\alpha_1, \ldots, \alpha_n) = \sum_{x=1}^{P} e^{2\pi i f(x)}.$$

Then under any positive integers $r \leqslant n$ and k we have the estimate

$$|S(\alpha_1, \ldots, \alpha_n)|^{2k} \leqslant k^{n-1} P^{\frac{n(n+1)}{2} - r} \sum_{|\lambda_r| < kP^r} N_k^{(P)}(0, \ldots, \lambda_r, \ldots, 0) e^{2\pi i \alpha_r \lambda_r}.$$

Proof. Let us consider the function

$$F(\alpha_1,\ldots,\alpha_n) = \left|S(\alpha_1,\ldots,\alpha_n)\right|^{2k}.$$

By (159)

$$F(\alpha_1,\ldots,\alpha_n) = \sum_{\lambda_1,\ldots,\lambda_n} N_k^{(P)}(\lambda_1,\ldots,\lambda_n)e^{2\pi i(\alpha_1\lambda_1+\ldots+\alpha_n\lambda_n)},$$

where the range of summation is

$$\left|\lambda_\nu\right| < kP^\nu \qquad (\nu = 1,2,\ldots,n). \tag{198}$$

Since $F(\alpha_1,\ldots,\alpha_n) \geqslant 0$, the lemma conditions are satisfied. Choose

$$q_\nu = \begin{cases} 1 & \text{if} \quad \nu = r, \\ kP^\nu & \text{if} \quad \nu \neq r. \end{cases} \tag{199}$$

Then using the lemma we get

$$\begin{aligned} &\left|S(\alpha_1,\ldots,\alpha_n)\right|^{2k} \\ &\qquad \leqslant q_1\cdots q_n \sum_{\lambda_1,\ldots,\lambda_n} N_k^{(P)}(q_1\lambda_1,\ldots,q_n\lambda_n)e^{2\pi i(\alpha_1 q_1\lambda_1+\ldots+\alpha_n q_n\lambda_n)}, \end{aligned} \tag{200}$$

where by (198) and (199) the range of summation may be written in the form

$$\left|\lambda_\nu\right| < \begin{cases} kP^r & \text{if} \quad \nu = r, \\ 1 & \text{if} \quad \nu \neq r, \end{cases}$$

or, that is just the same, in the form

$$\lambda_1 = \ldots = \lambda_{r-1} = 0, \qquad \left|\lambda_r\right| < kP^r, \qquad \lambda_{r+1} = \ldots = \lambda_n = 0.$$

Observing that

$$q_1\cdots q_n = k^{n-1}P^{\frac{n(n+1)}{2}-r},$$

from (200) we obtain the corollary assertion:

$$\begin{aligned} \left|S(\alpha_1,\ldots,\alpha_n)\right|^{2k} &\leqslant q_1\cdots q_n \sum_{|\lambda_r|<kP^r} N_k^{(P)}(0,\ldots,\lambda_r,\ldots,0)e^{2\pi i\,\alpha_r\lambda_r} \\ &= k^{n-1}P^{\frac{n(n+1)}{2}-r} \sum_{|\lambda_r|<kP^r} N_k^{(P)}(0,\ldots,\lambda_r,\ldots,0)e^{2\pi i\,\alpha_r\lambda_r}. \end{aligned}$$

THEOREM 17. *Let* $n > 2$, $f(x) = \alpha_1 x + \ldots + \alpha_n x^n$, *and*

$$\alpha_n = \frac{a}{q} + \frac{\theta}{q^2}, \qquad (a,q) = 1, \quad |\theta| \leqslant 1.$$

If $P \leqslant q \leqslant P^{n-1}$ *then the estimate*

$$\left|\sum_{x=1}^{P} e^{2\pi i f(x)}\right| \leqslant e^{3n}P^{1-\frac{1}{9n^2\log n}}$$

holds.

Proof. Using Lemma 19, we get

$$\left|\sum_{x=1}^{P} e^{2\pi i f(x)}\right|^{2k+1} \leqslant 2^{2k+1} \sum_{y=0}^{P-1} \left|\sum_{x=1}^{P} e^{2\pi i f(x+y)}\right|^{2k} = 2^{2k+1} \sum_{y=0}^{P-1} \left|\sum_{x=1}^{P} e^{2\pi i (\alpha_1(y)x+\ldots+\alpha_n(y)x^n)}\right|^{2k}, \tag{201}$$

where $\alpha_\nu(y) = \frac{1}{\nu!} f^{(\nu)}(y)$ and, in particular,

$$\alpha_{n-1}(y) = \frac{1}{(n-1)!} f^{(n-1)}(y) = n\alpha_n y + \alpha_{n-1}.$$

By virtue of the corollary of Lemma 20 we have

$$\left|\sum_{x=1}^{P} e^{2\pi i (\alpha_1(y)x+\ldots+\alpha_n(y)x^n)}\right|^{2k} \leqslant k^{n-1} P^{\frac{n(n-1)}{2}+1} \sum_{|\lambda_{n-1}|<kP^{n-1}} N_k^{(P)}(0,\ldots,\lambda_{n-1}0) e^{2\pi i (n\alpha_n y+\alpha_{n-1})\lambda_{n-1}}.$$

Substituting this estimate into (201), after interchanging the order of summation we obtain

$$\left|\sum_{x=1}^{P} e^{2\pi i f(x)}\right|^{2k+1} \leqslant k^{n-1} 2^{2k+1} P^{\frac{n(n-1)}{2}+1} \times \sum_{|\lambda_{n-1}|<kP^{n-1}} N_k^{(P)}(0,\ldots,\lambda_{n-1}0) \left|\sum_{y=0}^{P-1} e^{2\pi i n\alpha_n \lambda_{n-1} y}\right| \leqslant k^{n-1} 2^{2k+1} P^{\frac{n(n-1)}{2}+1} N_k(P) \sum_{|\lambda_{n-1}|<kP^{n-1}} \min\left(P, \frac{1}{2\|n\alpha_n\lambda_{n-1}\|}\right).$$

Choose $\varepsilon = \frac{1}{2n^2}$ in the note of Lemma 14 and use the condition $P \leqslant q \leqslant P^{n-1}$. Then we get

$$\sum_{|\lambda_{n-1}|<kP^{n-1}} \min\left(P, \frac{1}{2\|n\alpha_n\lambda_{n-1}\|}\right) \leqslant \frac{8n}{\varepsilon}\left(1+\frac{kP^{n-1}}{q}\right)(P+q)P^{\varepsilon} \leqslant 64 n^3 k P^{n-1+\frac{1}{2n^2}},$$

and, therefore,

$$\left|\sum_{x=1}^{P} e^{2\pi i f(x)}\right|^{2k+1} \leqslant n^3 k^n 2^{2k+7} P^{\frac{n(n+1)}{2}+\frac{1}{2n^2}} N_k(P). \tag{202}$$

Choose $\tau = [3n \log n] + 1$ in Theorem 16. Since, obviously,

$$\frac{n(n-1)}{2}\left(1-\frac{1}{n}\right)^{\tau} < \frac{n(n-1)}{2n^3} = \frac{1}{2n} - \frac{1}{2n^2},$$

then by (190) under $k = \frac{n(n+1)}{2} + n\tau$ we have

$$N_k(P) \leqslant (2\,k)^{2k}(2\,n)^{n^3} P^{2k-\frac{n(n+1)}{2}+\frac{1}{2n}-\frac{1}{2n^2}}.$$

Substituting this estimate into (202) and using $3n^2 \log n \leqslant k \leqslant 4n^2 \log n$ we obtain the theorem assertion:

$$\left|\sum_{x=1}^{P} e^{2\pi i f(x)}\right|^{2k+1} \leqslant 128\, n^3 k^{2k+n} 2^{4k} (2n)^{n^3} P^{2k+\frac{1}{2n}}$$

$$\leqslant e^{3n(2k+1)} P^{2k+1-\left(1-\frac{1}{2n}\right)},$$

$$\sum_{x=1}^{P} e^{2\pi i f(x)} \leqslant e^{3n} P^{1-\frac{1-\frac{1}{2n}}{2k+1}} \leqslant e^{3n} P^{1-\frac{1}{9n^2 \log n}}.$$

The Weyl's sums estimate obtained in Theorem 17 depends on rational approximations for the leading coefficient of the polynomial $f(x) = \alpha_1 x + \ldots + \alpha_n x^n$. Let us show that a similar estimate is valid in the case, when rational approximations of an arbitrary coefficient α_r $(2 \leqslant r \leqslant n)$ are given.

LEMMA 21. *Let Q, t be positive integers,*

$$\alpha = \frac{a}{q} + \frac{\theta}{q^2}, \qquad (a, q) = 1, \quad |\theta| \leqslant 1,$$

and T be the number of solutions of the inequality

$$\|\alpha x\| < \frac{t}{q}, \qquad |x| \leqslant Q.$$

Then we have the estimate

$$T \leqslant 6\left(1 + \frac{Q}{q}\right)t.$$

Proof. Denote by $T(\beta)$ the number of solutions of the inequality

$$\|\alpha x+\beta\|<\frac{t}{q}, \qquad 1\leqslant x\leqslant q.$$

By (149) under a certain integer b only depending on β, the estimate

$$\left\|\frac{ax+b}{q}\right\|\leqslant\|\alpha x+\beta\|+\frac{1}{q}$$

holds. Hence it follows that $T(\beta)$ does not exceed the number of solutions of the inequality

$$\left\|\frac{ax+b}{q}\right\|<\frac{t+1}{q}, \qquad 1\leqslant x\leqslant q.$$

Since $(a,q)=1$, the number of solutions of this inequality is equal to $2t+1$ and $T(\beta)\leqslant 2t+1$.

Consider now the inequality

$$\|\alpha x\|<\frac{t}{q}, \qquad -Q_1q<x\leqslant Q_1q, \tag{203}$$

where $Q_1=[\frac{Q}{q}]+1$. Replacing x by qx_1+x_2, we rewrite this inequality in the form

$$\|\alpha x_2+aqx_1\|<\frac{t}{q}, \qquad -Q_1\leqslant x_1\leqslant Q_1-1, \quad 1\leqslant x_2\leqslant q.$$

Since $Q_1q>Q$, then T does not exceed the number of solutions of the inequality (203) and, therefore,

$$T\leqslant\sum_{x_1=-Q_1}^{Q_1-1}T(\alpha qx_1).$$

Hence using the estimate $T(\alpha qx_1)\leqslant 2t+1$, we get the lemma assertion:

$$T\leqslant 2(2t+1)Q_1\leqslant 6t\left(1+\frac{Q}{q}\right).$$

LEMMA 22. *Let $n>2$, $f(x)=\alpha_1x+\ldots+\alpha_nx^n$, and*

$$\beta_s(x)=\frac{1}{s!}f^{(s)}(x)=\alpha_s+C_{s+1}^1\alpha_{s+1}x+\ldots+C_n^{n-s}\alpha_nx^{n-s} \qquad (s\geqslant 1).$$

If for some integers y, z, and t $(0\leqslant y,z<P)$ under $s=1,2,\ldots,n-1$ the inequalities

$$\|\beta_s(y)-\beta_s(z)\|<\frac{t}{P^s} \tag{204}$$

hold, then the inequalities

$$\|n!\,\alpha_{s+1}(y-z)\|<(2n)^{2n}\frac{t}{P^s} \qquad (s=1,2,\ldots,n-1)$$

are valid also.

Proof. Determine quantities h_s and γ_s by means of the equalities

$$h_s = \frac{n!}{(n-s+1)!}, \qquad \gamma_s = h_{s+1}\alpha_{n-s+1} \qquad (1 \leqslant s \leqslant n-1).$$

Then, obviously,

$$\beta_{n-s}(x) = \alpha_{n-s} + (n-s+1)\alpha_{n-s+1}x + \sum_{j=1}^{s-1} C_{n-j+1}^{s-j+1}\alpha_{n-j+1}x^{s-j+1},$$

$$h_s\beta_{n-s}(x) = h_s\alpha_{n-s} + \gamma_s x + \sum_{j=1}^{s-1} C_{n-j+1}^{s-j+1}\frac{h_s}{h_{j+1}}\gamma_j x^{s-j+1}$$

$$= h_s\alpha_{n-s} + \gamma_s x + \sum_{j=1}^{s-1} H_{sj}\gamma_j x^{s-j+1}, \tag{205}$$

where the quantities H_{sj} are determined by the equalities

$$H_{sj} = C_{n-j+1}^{s-j+1}\frac{h_s}{h_{j+1}} = C_{n-j+1}^{s-j+1}\frac{(n-j)!}{(n-s+1)!} \qquad (1 \leqslant j \leqslant s-1).$$

It is seen from the definition of h_s and H_{sj}, that the estimates

$$h_s \leqslant n^{s-1}, \qquad H_{sj} \leqslant n^{2s-2j} \tag{206}$$

hold, and that h_s and H_{sj} are integers.

Using the fact that under an integer m and an arbitrary γ the estimate $\|m\gamma\| \leqslant |m|\,\|\gamma\|$ is valid, we obtain from (205)

$$h_s\big(\beta_{n-s}(y) - \beta_{n-s}(z)\big) = \gamma_s(y-z) + \sum_{j=1}^{s-1} H_{sj}\gamma_j\big(y^{s-j+1} - z^{s-j+1}\big),$$

$$\|\gamma_s(y-z)\| \leqslant \big\|h_s\big(\beta_{n-s}(y) - \beta_{n-s}(z)\big)\big\| + \sum_{j=1}^{s-1}\big\|H_{sj}\gamma_j(y-z)\big(y^{s-j} + \ldots + z^{s-j}\big)\big\|$$

$$\leqslant h_s\big\|\beta_{n-s}(y) - \beta_{n-s}(z)\big\| + \sum_{j=1}^{s-1} H_{sj}\big(y^{s-j} + \ldots + z^{s-j}\big)\|\gamma_j(y-z)\|. \tag{207}$$

Now we shall show that under $s = 1, 2, \ldots, n-1$ the inequality

$$\|\gamma_s(y-z)\| < (2n)^{2s-2}\frac{t}{P^{n-s}} \tag{208}$$

holds.

Indeed, under $s = 1$ this inequality coincides with the last of the inequalities (204):

$$\|\gamma_1(y-z)\| = \|\beta_{n-1}(y) - \beta_{n-1}(z)\| < \frac{t}{P^{n-1}}.$$

Apply the induction. Let $2 \leqslant s \leqslant n-1$ and the inequality (208) be fulfilled under $j \leqslant s-1$:

$$\|\gamma_j(y-z)\| \leqslant (2n)^{2j-2}\frac{t}{P^{n-j}} \qquad (j = 1, 2, \ldots, s-1). \tag{209}$$

If follows from (207) by virtue of (206) that

$$\|\gamma_s(y-z)\| \leqslant n^{s-1}\|\beta_{n-s}(y) - \beta_{n-s}(z)\| + \sum_{j=1}^{s-1} n^{2s-2j}(s-j+1)P^{s-j}\|\gamma_j(y-z)\|.$$

Hence, using the induction hypothesis and the estimate (204), we get

$$\|\gamma_s(y-z)\| \leqslant n^{s-1}\frac{t}{P^{n-s}} + \sum_{j=1}^{s-1} n^{2s-2j}(s-j+1)P^{s-j}(2n)^{2j-2}\frac{t}{P^{n-j}}$$
$$\leqslant \left(n^{s-1} + n^{2s-2}\sum_{j=1}^{s-1}(s-j+1)2^{2j-2}\right)\frac{t}{P^{n-s}} < (2n)^{2s-2}\frac{t}{P^{n-s}}.$$

Thus, if the inequalities (209) hold under $j \leqslant s-1$, then they hold under $j \leqslant s$ too. Therefore these inequalities hold under any $j \leqslant n-1$. This coincides with the assertion (208).

Now, observing that $n!\,\alpha_{s+1} = s!\gamma_{n-s}$ and using the estimate (208), we get the lemma assertion:

$$\|n!\,\alpha_{s+1}(y-z)\| = \|s!\gamma_{n-s}(y-z)\| \leqslant s!\|\gamma_{n-s}(y-z)\|$$
$$\leqslant s!(2n)^{2n-2s-2}\frac{t}{P^s} < (2n)^{2n}\frac{t}{P^s} \qquad (s = 1, 2, \ldots, n-1).$$

LEMMA 23. *Let $n > 2$, $f(x) = \alpha_1 x + \ldots + \alpha_n x^n$, $\beta_s(x) = \frac{1}{s!}f^{(s)}(x)$, and under a certain r from the interval $2 \leqslant r \leqslant n$*

$$\alpha_r = \frac{a}{q} + \frac{\theta}{q^2}, \qquad (a, q) = 1, \quad |\theta| \leqslant 1.$$

Further let $P \leqslant q \leqslant P^{r-1}$ and the sum

$$\sum_{y,z}{}_1 = \sum_{y,z}{}_1 \prod_{\nu=1}^{n-1} \min\left(P^\nu, \frac{1}{\|\beta_\nu(y) - \beta_\nu(z)\|}\right)$$

be extended over those values of y and z ($0 \leqslant y, z < P$), which under a certain positive integer t satisfy the inequalities

$$\|\beta_s(y) - \beta_s(z)\| < \frac{t}{P^s} \qquad (s = 1, 2, \ldots, n-1). \tag{210}$$

Then we have the estimate

$$\sum_{y,z} {}_1 \leqslant (2n)^{3n} P^{\frac{n(n-1)}{2}+1} t.$$

Proof. Denote by T_1 the number of summands in the sum $\sum_1$. Then estimating all the summands trivially, we obtain

$$\sum_{y,z} {}_1 \leqslant P^{\frac{n(n-1)}{2}} T_1. \tag{211}$$

Since T_1 is the number of those values of y and z, which satisfy the conditions (210), then by Lemma 22, T_1 does not exceed the number of solutions of the system of inequalities

$$\|n!\,\alpha_{s+1}(y-z)\| < (2n)^{2n}\frac{t}{P^s}, \qquad 0 \leqslant y, z < P \quad (s = 1, 2, \ldots, n-1),$$

and, therefore, does not exceed the number of solutions of the inequality

$$\|n!\,\alpha_r(y-z)\| < (2n)^{2n}\frac{t}{P^{r-1}}, \qquad 0 \leqslant y, z < P.$$

Replace $n!\,(y-z)$ by x. Then, obviously, $|x| < n!P$ and the quantity x takes on each integer value at most P times. Therefore $T_1 \leqslant PT$, where T is the number of solutions of the inequality

$$\|\alpha_r x\| < (2n)^{2n}\frac{t}{P^{r-1}}, \qquad |x| < n!P.$$

Since $q \leqslant P^{r-1}$, then T does not exceed the number of solutions of the inequality

$$\|\alpha_r x\| < (2n)^{2n}\frac{t}{q}, \qquad |x| < n!P,$$

and by Lemma 21

$$T \leqslant 6\left(1 + \frac{n!P}{q}\right)(2n)^{2n}t.$$

But then, observing that $q \geqslant P$, we get

$$T_1 \leqslant 6\left(1+\frac{n!P}{q}\right)(2n)^{2n}tP \leqslant (2n)^{3n}P\,t.$$

The lemma assertion follows by substituting this estimate into (211):

$$\sum_{y,z} 1 \leqslant P^{\frac{n(n-1)}{2}} T_1 \leqslant (2n)^{3n} P^{\frac{n(n-1)}{2}+1} t.$$

THEOREM 18. *Let* $n > 2$, $f(x) = \alpha_1 x + \ldots + \alpha_n x^n$, $2 \leqslant r \leqslant n$, *and*

$$\alpha_r = \frac{a}{q} + \frac{\theta}{q^2}, \qquad (a, q) = 1, \quad |\theta| \leqslant 1.$$

If $P \leqslant q \leqslant P^{r-1}$, *then*

$$\left|\sum_{x=1}^{P} e^{2\pi i\, f(x)}\right| \leqslant e^{3n} P^{1-\frac{1}{24\, n^2 \log\, n}}.$$

Proof. Determine quantities $\beta_s(y)$ with the help of the equalities

$$\beta_s(y) = \frac{1}{s!}\, f^{(s)}(y) \qquad (s = 0, 1, \ldots, n).$$

Then, obviously,

$$f(x+y) = \beta_0(y) + \beta_1(y)x + \ldots + \beta_n(y)x^n$$

and according to Lemma 19

$$\left|\sum_{x=1}^{P} e^{2\pi i\, f(x)}\right|^{2k+1} \leqslant 2^{2k+1} \sum_{y=0}^{P-1} \left|\sum_{x=1}^{P} e^{2\pi i\, f(x+y)}\right|^{2k}$$

$$= 2^{2k+1} \sum_{y=0}^{P-1} \left|\sum_{x=1}^{P} e^{2\pi i\,(\beta_1(y)x + \ldots + \beta_n(y)x^n)}\right|^{2k}.$$

Further, using the equality (159), we obtain

$$\left|\sum_{x=1}^{P} e^{2\pi i\, f(x)}\right|^{2k+1} \leqslant 2^{2k+1} \sum_{y=0}^{P-1} \sum_{\lambda_1,\ldots,\lambda_n} N_k^{(P)}(\lambda_1,\ldots,\lambda_n) e^{2\pi i\,(\lambda_1\beta_1(y)+\ldots+\lambda_n\beta_n(y))}$$

$$\leqslant 2^{2k+1} \sum_{\lambda_1,\ldots,\lambda_n} N_k^{(P)}(\lambda_1,\ldots,\lambda_n) \left|\sum_{y=0}^{P-1} e^{2\pi i\,(\lambda_1\beta_1(y)+\ldots+\lambda_n\beta_n(y))}\right|,$$

where the range of summation is

$$|\lambda_\nu| < kP^\nu \qquad (\nu = 1, 2, \ldots, n).$$

Hence, using the Cauchy inequality (143) and the relation (165), we get

$$\begin{aligned}\left|\sum_{x=1}^{P} e^{2\pi i\, f(x)}\right|^{4k+2} &\leqslant 2^{4k+2} \sum_{\lambda_1,\ldots,\lambda_n} \left[N_k^{(P)}(\lambda_1,\ldots,\lambda_n)\right]^2 \sum_{\lambda_1,\ldots,\lambda_n} \left|\sum_{y=0}^{P-1} e^{2\pi i\,(\lambda_1\beta_1(y)+\ldots+\lambda_n\beta_n(y))}\right|^2 \\ &= 2^{4k+2} N_{2k}(P)V(P),\end{aligned} \tag{212}$$

where

$$V(P) = \sum_{\lambda_1,\ldots,\lambda_n} \left|\sum_{y=0}^{P-1} e^{2\pi i\,(\lambda_1\beta_1(y)+\ldots+\lambda_n\beta_n(y))}\right|^2.$$

Now we shall estimate the magnitude of $V(P)$. Observing that $\beta_n(y)$ does not depend on y, we obtain

$$\begin{aligned} V(P) &\leqslant 2kP^n \sum_{y,z=0}^{P-1} \sum_{\lambda_1,\ldots,\lambda_{n-1}} e^{2\pi i\,((\beta_1(y)-\beta_1(z))\lambda_1+\ldots+(\beta_{n-1}(y)-\beta_{n-1}(z))\lambda_{n-1})} \\ &\leqslant 2kP^n \sum_{y,z=0}^{P-1} \sum_{\nu=1}^{n-1} \min\left(2kP^\nu, \frac{1}{2\|\beta_\nu(y)-\beta_\nu(z)\|}\right) \\ &\leqslant (2k)^n P^n \left(\sum_{y,z}{}_1 + \sum_{y,z}{}_2\right), \end{aligned} \tag{213}$$

where the sum

$$\sum_{y,z}{}_1 = \sum_{y,z}{}_1 \prod_{\nu=1}^{n-1} \min\left(P^\nu, \frac{1}{\|\beta_\nu(y)-\beta_\nu(z)\|}\right)$$

is extended over those values of y and z, which under a certain $t \leqslant 1$ satisfy the inequalities

$$\|\beta_s(y)-\beta_s(z)\| < \frac{t}{P^s} \qquad (s = 1, 2, \ldots, n-1).$$

Respectively, the sum

$$\sum_{y,z}{}_2 = \sum_{y,z}{}_2 \prod_{\nu=1}^{n-1} \min\left(P^\nu, \frac{1}{\|\beta_\nu(y)-\beta_\nu(z)\|}\right) \tag{214}$$

is over those values of y and z, for which there is s $(1 \leqslant s \leqslant n-1)$ such that

$$\|\beta_s(y)-\beta_s(z)\| \geqslant \frac{t}{P^s}. \tag{215}$$

By Lemma 23 for the sum $\sum_1$ the estimate

$$\sum_{y,z}{}_1 \leqslant (2n)^{3n} P^{\frac{n(n-1)}{2}+1} t$$

holds. Applying the estimate (215) for one of factors in (214) and estimating all the other factors trivially, we get

$$\sum_{y,z}{}_2 \leqslant \sum_{y,z=0}^{P-1} P^{\frac{n(n-1)}{2}} \frac{1}{t} = \frac{1}{t} P^{\frac{n(n-1)}{2}+2}.$$

Since by virtue of (213)

$$V(P) \leqslant (2k)^n P^n \Big(\sum_{y,z}{}_1 + \sum_{y,z}{}_2\Big),$$

then choosing $t=[\sqrt{P}]+1$ we obtain

$$\begin{aligned} V(P) &\leqslant (2k)^n P^n \Big((2n)^{3n} P^{\frac{n(n-1)}{2}+1} 2\sqrt{P} + \frac{1}{\sqrt{P}} P^{\frac{n(n-1)}{2}+2}\Big) \\ &\leqslant 3(2k)^n (2n)^{3n} P^{\frac{n(n+1)}{2}+\frac{3}{2}}. \end{aligned}$$

Substituting this estimate into (212), we get

$$\left|\sum_{x=1}^{P} e^{2\pi i f(x)}\right|^{4k+2} \leqslant 3(2k)^n 2^{4k+2} (2n)^{3n} N_{2k}(P) P^{\frac{n(n+1)}{2}+\frac{3}{2}}. \tag{216}$$

Choose

$$\begin{aligned} \tau_1 &= \Big[\frac{n}{2}+\frac{3}{2} n \log n\Big]+1, \qquad \tau = 2\tau_1, \\ k &= \Big[\frac{1}{2}+\frac{n(n+1)}{4}\Big]+n\tau_1. \end{aligned}$$

It is easy to verify that the estimates

$$2k \geqslant \frac{n(n+1)}{2}+n\tau, \qquad \tau > 3n\log n + n,$$

$$\frac{n(n-1)}{2}\left(1-\frac{1}{n}\right)^{\tau} < \frac{n(n-1)}{2en^3} < \frac{1}{24}$$

hold. Therefore, using Theorem 16, we obtain

$$N_{2k}(P) \leqslant (4k)^{4k}(2n)^{n^3} P^{4k-\frac{n(n+1)}{2}+\frac{1}{24}}.$$

The theorem follows by substituting this estimate into (216):

$$\left|\sum_{x=1}^{P} e^{2\pi i\, f(x)}\right|^{4k+2} \leqslant (8k)^{4k+2}(2n)^{n^3+3n}(2k)^n P^{4k+\frac{37}{24}}$$

$$\leqslant e^{3n(4k+2)} P^{4k+2-\frac{11}{24}},$$

$$\left|\sum_{x=1}^{P} e^{2\pi i\, f(x)}\right| \leqslant e^{3n} P^{1-\frac{11}{24(4k+2)}} \leqslant e^{3n} P^{1-\frac{1}{24n^2 \log n}}.$$

The estimates of the form

$$\left|\sum_{x=1}^{P} e^{2\pi i\, f(x)}\right| \leqslant C(n) P^{1-\frac{\gamma}{n^2 \log n}} \tag{217}$$

obtained in Theorems 17 and 18 are established on the assumption that $P \leqslant q \leqslant P^{r-1}$, where q is the denominator of rational approximations of the r-th coefficient of the polynomial $f(x) = \alpha_1 x + \ldots + \alpha_n x^n$:

$$\alpha_r = \frac{a}{q} + \frac{\theta}{q^2}, \qquad (a, q) = 1, \quad |\theta| \leqslant 1 \quad (2 \leqslant r \leqslant n).$$

It can be shown that the estimate (217) holds under $P^{\varepsilon} \leqslant q \leqslant P^{r-\varepsilon}$ with an arbitrary $\varepsilon > 0$ too, but it leads, as in the Weyl method (see the note of Theorem 14), to worsening the constant γ.

§ 15. Repeated application of the mean value theorem

Let $f(x) = \alpha_1 x + \ldots + \alpha_{n+1} x^{n+1}$ and $S(P)$ be Weyl's sum

$$S(P) = \sum_{x=1}^{P} e^{2\pi i\, f(x)}. \tag{218}$$

We shall write the estimate for this sum in the form

$$|S(P)| \leqslant C(n) P^{1-\rho} \tag{219}$$

and call P^ρ *a reducing factor*.

A generic peculiarity of different methods of the estimation of Weyl's sums consists in the fact that a reducing factor becomes smaller as in the process of obtaining the estimate (219) the sum $S(P)$ is raised to a greater power. So in the methods of Mordell, Vinogradov, and Weyl the sum is raised to a power having the order n, $n^2 \log n$ and 2^n, respectively, ultimately it leads to estimates with reducing factors $P^{\frac{1}{n}}$, $P^{\frac{\gamma_1}{n^2 \log n}}$ and $P^{\frac{\gamma_2}{2^n}}$, where γ_1 and γ_2 are certain positive constants.

Results exposed in this section are of another character. Here Weyl's sum is raised to a comparatively large power having the order up to n^4, however this does not lead to worsening estimates. On the contrary, it becomes possible to improve the reducing factor and besides to decrease (or even replace by an absolute constant) the coefficient $C(n)$ in the estimate (219). The last circumstance is of great importance in those cases, when under the growth of P the degree of the polynomial $f(x)$ grows also. These results are based upon the following lemma.

LEMMA 24. *Under any positive integers k_1 and k_2 for the sum (218) we have the estimate*

$$\left|S(P)\right|^{4k_1k_2+2k_2} \leqslant 2^{4k_1k_2+4k_2} P^{4k_1k_2-2k_1} V,$$

where

$$V = \Sigma N_{k_1}^{(P)}(\lambda_1, \ldots, \lambda_n) N_{k_2}^{(P)}(\mu_1, \ldots, \mu_n) e^{2\pi i(\beta_1\mu_1 + \ldots + \beta_n\mu_n)}, \tag{220}$$

the range of summation is

$$\left|\lambda_\nu\right| < k_1 P^\nu, \qquad \left|\mu_\nu\right| < k_2 P^\nu \qquad (\nu = 1, 2, \ldots, n)$$

and the quantities β_ν are determined under $1 \leqslant \nu \leqslant n$ by the equality

$$\beta_\nu = C_{\nu+1}^\nu \alpha_{\nu+1}\lambda_1 + \ldots + C_{n+1}^\nu \alpha_{n+1}\lambda_{n+1-\nu}.$$

Proof. Consider the sum

$$S_1 = \sum_{y=1}^{P} \left| \sum_{x=1}^{P} e^{2\pi i f(x+y)} \right|^{2k_1}.$$

Define quantities $\alpha_\nu(y)$ and β_0 with the help of the equalities

$$\alpha_\nu(y) = \frac{1}{\nu!} f^{(\nu)}(y) \quad (\nu = 0, 1, \ldots, n+1),$$
$$\beta_0 = \alpha_1\lambda_1 + \ldots + \alpha_n\lambda_n$$

and write the polynomial $f(x+y)$ in the form

$$f(x+y) = \alpha_0(y) + \alpha_1(y)x + \ldots + \alpha_{n+1}(y)x^{n+1}.$$

By (159)

$$\left|\sum_{x=1}^{P} e^{2\pi i(\alpha_1(y)x+\ldots+\alpha_{n+1}(y)x^{n+1})}\right|^{2k_1} = \sum_{\lambda_1,\ldots,\lambda_{n+1}} N_{k_1}^{(P)}(\lambda_1,\ldots,\lambda_{n+1}) e^{2\pi i(\alpha_1(y)\lambda_1+\ldots+\alpha_{n+1}(y)\lambda_{n+1})}$$

and, therefore,

$$S_1 = \sum_{y=1}^{P}\left|\sum_{x=1}^{P} e^{2\pi i(\alpha_1(y)x+\ldots+\alpha_{n+1}(y)x^{n+1})}\right|^{2k_1} = \sum_{\lambda_1,\ldots,\lambda_{n+1}} N_{k_1}^{(P)}(\lambda_1,\ldots,\lambda_{n+1}) \sum_{y=1}^{P} e^{2\pi i(\alpha_1(y)\lambda_1+\ldots+\alpha_{n+1}(y)\lambda_{n+1})}.$$

Since by (163)

$$\sum_{\lambda_{n+1}} N_{k_1}^{(P)}(\lambda_1,\ldots,\lambda_{n+1}) = N_{k_1}^{(P)}(\lambda_1,\ldots,\lambda_n),$$

then observing that $\alpha_{n+1}(y)$ does not depend on y, we get

$$S_1 \leqslant \sum_{\lambda_1,\ldots,\lambda_{n+1}} N_{k_1}^{(P)}(\lambda_1,\ldots,\lambda_{n+1})\left|\sum_{y=1}^{P} e^{2\pi i(\alpha_1(y)\lambda_1+\ldots+\alpha_n(y)\lambda_n)}\right| = \sum_{\lambda_1,\ldots,\lambda_n} N_{k_1}^{(P)}(\lambda_1,\ldots,\lambda_n)\left|\sum_{y=1}^{P} e^{2\pi i(\alpha_1(y)\lambda_1+\ldots+\alpha_n(y)\lambda_n)}\right|.$$

Applying the inequality (141) and using the relation

$$\alpha_1(y)\lambda_1 + \ldots + \alpha_n(y)\lambda_n = \beta_0 + \beta_1 y + \ldots + \beta_n y^n,$$

which follows from the definition of the quantities β_ν and $\alpha_\nu(y)$, we obtain

$$\begin{aligned} S_1^{2k_2} &\leqslant \left(\sum_{\lambda_1,\ldots,\lambda_n} N_{k_1}^{(P)}(\lambda_1,\ldots,\lambda_n)\right)^{2k_2-1} \\ &\times \sum_{\lambda_1,\ldots,\lambda_n} N_{k_1}^{(P)}(\lambda_1,\ldots,\lambda_n)\left|\sum_{y=1}^{P} e^{2\pi i(\beta_1 y+\ldots+\beta_n y^n)}\right|^{2k_2} \\ &= P^{2k_1(2k_2-1)} \sum_{\lambda_1,\ldots,\lambda_n} N_{k_1}^{(P)}(\lambda_1,\ldots,\lambda_n)\left|\sum_{y=1}^{P} e^{2\pi i(\beta_1 y+\ldots+\beta_n y^n)}\right|^{2k_2} \\ &= P^{2k_1(2k_2-1)} V, \end{aligned} \tag{221}$$

where V is determined by the equality (220).

Let us show that $V \geqslant P^{2k_1}$. Indeed, it is seen from the determination of the quantities β_ν that in the equality (220) the sum $\beta_1\mu_1 + \ldots + \beta_n\mu_n$ is a homogeneous linear function of the quantities $\lambda_1, \ldots, \lambda_n$:

$$\beta_1\mu_1 + \ldots + \beta_n\mu_n = \gamma_1\lambda_1 + \ldots + \gamma_n\lambda_n.$$

But then by (159)

$$\sum_{\lambda_1,\ldots,\lambda_n} N_{k_1}^{(P)}(\lambda_1, \ldots, \lambda_n) e^{2\pi i(\beta_1\mu_1+\ldots+\beta_n\mu_n)} = \sum_{\lambda_1,\ldots,\lambda_n} N_{k_1}^{(P)}(\lambda_1, \ldots, \lambda_n) e^{2\pi i(\gamma_1\lambda_1+\ldots+\gamma_n\lambda_n)} \geqslant 0, \tag{222}$$

and the quantity V can be estimated by the summand obtained under $\mu_1 = \ldots = \mu_n = 0$:

$$V = \sum_{\mu_1,\ldots,\mu_n} N_{k_2}^{(P)}(\mu_1, \ldots, \mu_n) \sum_{\lambda_1,\ldots,\lambda_n} N_{k_1}^{(P)}(\lambda_1, \ldots, \lambda_n) e^{2\pi i(\beta_1\mu_1+\ldots+\beta_n\mu_n)} \geqslant N_{k_2}(P) \sum_{\lambda_1,\ldots,\lambda_n} N_{k_1}^{(P)}(\lambda_1, \ldots, \lambda_n) = P^{2k_1} N_{k_2}(P) \geqslant P^{2k_1}. \tag{223}$$

Now, using the inequality (195), we get

$$\begin{aligned} |S(P)|^{2k_1+1} &\leqslant 2^{2k_1+1} \sum_{y=0}^{P-1} \left| \sum_{x=1}^{P} e^{2\pi i f(x+y)} \right|^{2k_1} \\ &\leqslant 2^{2k_1+1} \sum_{y=0}^{P} \left| \sum_{x=1}^{P} e^{2\pi i f(x+y)} \right|^{2k_1} \\ &\leqslant 2^{2k_1+1} \left(P^{2k_1} + \sum_{y=1}^{P} \left| \sum_{x=1}^{P} e^{2\pi i f(x+y)} \right|^{2k_1} \right) = 2^{2k_1+1}(P^{2k_1} + S_1). \end{aligned}$$

Hence, using the inequalities (221) and (223), we obtain the lemma assertion:

$$\begin{aligned} |S(P)|^{4k_1k_2+2k_2} &\leqslant 2^{4k_1k_2+4k_2-1}(P^{4k_1k_2} + S_1^{2k_2}) \\ &\leqslant 2^{4k_1k_2+4k_2-1}(P^{4k_1k_2} + P^{4k_1k_2-2k_1}V) \leqslant 2^{4k_1k_2+4k_2} P^{4k_1k_2-2k_1} V. \end{aligned}$$

COROLLARY. *Under any positive integers k_1, k_2, and m $(1 \leqslant m \leqslant n)$ for the sum (218) we have the estimate*

$$|S(P)|^{4k_1k_2+2k_2} \leqslant (2k_2)^n \, 2^{4k_1k_2+4k_2} P^{4k_1k_2-2k_1+\frac{m(m-1)}{2}} N_{k_1,n+1-m}(P) N_{k_2,n}(P)\sigma, \tag{224}$$

where

$$\sigma = \sum_{\lambda_1,\ldots,\lambda_{n+1-m}} \min\left(P^m, \frac{1}{\|\beta_m\|}\right) \ldots \min\left(P^n, \frac{1}{\|\beta_n\|}\right), \tag{225}$$

the summation range is $|\lambda_j| < k_1 P^j$ *and the quantities* β_ν *are determined by the equalities*

$$\beta_\nu = C^\nu_{\nu+1}\alpha_{\nu+1}\lambda_1 + \ldots + C^\nu_{n+1}\alpha_{n+1}\lambda_{n+1-\nu} \qquad (1 \leqslant \nu \leqslant n).$$

Proof. Since by (222)

$$\sum_{\lambda_1,\ldots,\lambda_n} N^{(P)}_{k_1}(\lambda_1,\ldots,\lambda_n) e^{2\pi i(\beta_1\mu_1+\ldots+\beta_n\mu_n)} \geqslant 0,$$

then, obviously,

$$\begin{aligned} V &= \sum_{\mu_1,\ldots,\mu_n} N^{(P)}_{k_2}(\mu_1,\ldots,\mu_n) \sum_{\lambda_1,\ldots,\lambda_n} N^{(P)}_{k_1}(\lambda_1,\ldots,\lambda_n) e^{2\pi i(\beta_1\mu_1+\ldots+\beta_n\mu_n)} \\ &\leqslant N_{k_2,n}(P) \sum_{\mu_1,\ldots,\mu_n} \sum_{\lambda_1,\ldots,\lambda_n} N^{(P)}_{k_1}(\lambda_1,\ldots,\lambda_n) e^{2\pi i(\beta_1\mu_1+\ldots+\beta_n\mu_n)}. \end{aligned}$$

Interchanging the order of summation, we obtain

$$\begin{aligned} V &\leqslant N_{k_2,n}(P) \\ &\times \sum_{\lambda_1,\ldots,\lambda_n} N^{(P)}_{k_1}(\lambda_1,\ldots,\lambda_n) \min\left(2k_1P, \frac{1}{2\|\beta_1\|}\right) \ldots \min\left(2k_1P^n, \frac{1}{2\|\beta_n\|}\right) \\ &\leqslant (2k_1)^n P^{\frac{m(m-1)}{2}} N_{k_2,n}(P) \\ &\times \sum_{\lambda_1,\ldots,\lambda_n} N^{(P)}_{k_1}(\lambda_1,\ldots,\lambda_n) \min\left(P^m, \frac{1}{\|\beta_m\|}\right) \ldots \min\left(P^n, \frac{1}{\|\beta_n\|}\right). \end{aligned}$$

Hence, observing that under $m \geqslant 2$ the quantities $\beta_1,\ldots,\beta_m$ do not depend on $\lambda_{n+2-m},\ldots, \lambda_n$ and using the equality

$$\sum_{\lambda_{n+2-m},\ldots,\lambda_n} N^{(P)}_{k_1}(\lambda_1,\ldots,\lambda_n) = N^{(P)}_{k_1}(\lambda_1,\ldots,\lambda_{n+1-m}),$$

we get the corollary assertion:

$$\begin{aligned} V &\leqslant (2k_1)^n P^{\frac{m(m-1)}{2}} N_{k_2,n}(P) \\ &\times \sum_{\lambda_1,\ldots,\lambda_{n+1-m}} N^{(P)}_{k_1}(\lambda_1,\ldots,\lambda_{n+1-m}) \min\left(P^m, \frac{1}{\|\beta_m\|}\right) \ldots \min\left(P^n, \frac{1}{\|\beta_n\|}\right) \\ &\leqslant (2k_1)^n P^{\frac{m(m-1)}{2}} N_{k_1,n+1-m}(P) N_{k_2,n}(P)\sigma, \end{aligned}$$

$$|S(P)|^{4k_1k_2+2k_2} \leqslant 2^{4k_1k_2+4k_2} P^{4k_1k_2-2k_1} V$$
$$\leqslant (2k_1)^n 2^{4k_1k_2+4k_2} P^{4k_1k_2-2k_1+\frac{m(m-1)}{2}} N_{k_1,n+1-m}(P) N_{k_2,n}(P)\sigma,$$

where

$$\sigma = \sum_{\lambda_1,\ldots,\lambda_{n+1-m}} \min\left(P^m, \frac{1}{\|\beta_m\|}\right) \ldots \min\left(P^n, \frac{1}{\|\beta_n\|}\right).$$

It is seen from the inequality (224) that Lemma 24 reduces the estimation of the sum $S(P)$ to the estimation of the product of the quantities $N_{k_1,\,n+1-m}(P)$ and $N_{k_2,\,n}(P)$. In estimating of this product one has to apply the Vinogradov mean value theorem twice. That is why the use of Lemma 24 in estimating Weyl's sums is referred to as the repeated application of the mean value theorem. Let us show that the repeated application of the mean value theorem enables us to strengthen the estimates of Weyl's sums obtained in Theorem 17.

THEOREM 19. *Let* $P > 1$, $n > 2$, $f(x) = \alpha_1 x + \ldots + \alpha_{n+1} x^{n+1}$,

$$\alpha_{n+1} = \frac{a}{q} + \frac{\theta}{q^2}, \qquad (a, q) = 1, \quad |\theta| \leqslant 1,$$

and r *be determined by the equality* $q = P^r$. *Then under any* r *from the interval* $1 \leqslant r \leqslant n$ *we have the estimate*

$$\left|\sum_{x=1}^{P} e^{2\pi i f(x)}\right| \leqslant 2 e^{\frac{2n \log n}{s^2(\log 3n - \log s)}} P^{1-\frac{1}{95n^2(\log 3n - \log s)}}, \tag{226}$$

where $s = \min([r],\ [n+1-r])$.

Proof. According to the corollary of Lemma 24 the estimation of the sum (226) is reduced to the estimation of the magnitude of

$$\sigma = \sum_{\lambda_1,\ldots,\lambda_{n+1-m}} \min\left(P^m, \frac{1}{\|\beta_m\|}\right) \ldots \min\left(P^n, \frac{1}{\|\beta_n\|}\right),$$

where the summation is extended over the region $|\lambda_j| < k_1 P^j$ $(1 \leqslant j \leqslant n+1-m)$ and

$$\beta_\nu = C_{\nu+1}^{\nu} \alpha_{\nu+1} \lambda_1 + \ldots + C_{n+1}^{\nu} \alpha_{n+1} \lambda_{n+1-\nu}.$$

Determine β'_ν by means of the equality $\beta'_\nu = C_{\nu+1}^{\nu} \alpha_{\nu+1} \lambda_1 + \ldots + C_n^{\nu} \alpha_n \lambda_{n-\nu}$ and write β_ν in the form

$$\beta_\nu = C_{n+1}^{\nu} \alpha_{n+1} \lambda_{n+1-\nu} + \beta'_\nu \qquad (m \leqslant \nu \leqslant n).$$

Since β'_ν does not depend on $\lambda_{n+1-\nu}$, then by the note of Lemma 14 under $\varepsilon = \frac{1}{2n^2}$ we have the estimate

$$\sum_{\lambda_{n+1-\nu}} \min\left(P^\nu, \frac{1}{\|\beta_\nu\|}\right) = \sum_{|x|<k_1 P^{n+1-\nu}} \min\left(P^\nu, \frac{1}{\|C_{n+1}^\nu \alpha_{n+1} x + \beta'_\nu\|}\right)$$
$$\leqslant C_{n+1}^\nu \frac{8}{\varepsilon}\left(1 + \frac{k_1 P^{n+1-\nu}}{q}\right)(P^\nu + q)P^{\varepsilon\nu}$$
$$\leqslant k_1 2^{4n}\left(1 + \frac{P^{n+1-\nu}}{q}\right)(P^\nu + q)P^{\frac{1}{2n}}.$$

Denote by σ_1 the sum

$$\sigma_1 = \sum_{\lambda_1,\ldots,\lambda_{n-m}} \min\left(P^{m+1}, \frac{1}{\|\beta_{m+1}\|}\right)\ldots\min\left(P^n, \frac{1}{\|\beta_n\|}\right).$$

Observing that among the quantities $\beta_m, \ldots, \beta_n$ only β_m depends on λ_{n+1-m}, we get

$$\sigma = \sum_{\lambda_1,\ldots,\lambda_{n-m}} \prod_{\nu=m+1}^{n} \min\left(P^\nu, \frac{1}{\|\beta_\nu\|}\right) \sum_{\lambda_{n+1-m}} \min\left(P^m, \frac{1}{\|\beta_m\|}\right)$$
$$\leqslant k_1 2^{4n}\left(1 + \frac{P^{n+1-m}}{q}\right)(P^m + q)P^{\frac{1}{2n}}\sigma_1. \tag{227}$$

Further let the sum σ_2 be determined by the equality

$$\sigma_2 = \sum_{\lambda_1,\ldots,\lambda_{n-m-1}} \min\left(P^{m+2}, \frac{1}{\|\beta_{m+2}\|}\right)\ldots\min\left(P^n, \frac{1}{\|\beta_n\|}\right).$$

Since among the quantities $\beta_{m+1}, \ldots, \beta_n$ only β_{m+1} depends on λ_{n-m}, then similar to (227) we obtain

$$\sigma_1 \leqslant k_1 2^{4n}\left(1 + \frac{P^{n-m}}{q}\right)(P^{m+1} + q)P^{\frac{1}{2n}}\sigma_2.$$

Continuing this process, finally we arrive at the estimate

$$\sigma \leqslant k_1^{n+1-m} 2^{4n(n+1-m)} P^{\frac{n+1-m}{2n}} \prod_{\nu=m}^{n}\left(1 + \frac{P^{n+1-\nu}}{q}\right)(P^\nu + q).$$

Choose $m = n + 1 - s$. Then

$$m = n + 1 - \min\left([r], [n+1-r]\right) \geqslant r,$$
$$n + 1 - m = \min\left([r], [n+1-r]\right) \leqslant r,$$

and under $\nu \geqslant m$ the inequality $n+1-\nu \leqslant r \leqslant \nu$ is satisfied. Therefore, $P^{n+1-\nu} \leqslant q$, $P^{\nu} \geqslant q$, and

$$\sigma \leqslant k_1^n 2^{4n^2} P^{\frac{1}{2}} \prod_{\nu=m}^{n} 4P^{\nu} \leqslant (4k_1)^n 2^{4n^2} P^{\frac{n(n+1)}{2} - \frac{m(m-1)}{2} + \frac{1}{2}}.$$

But then according to the corollary of Lemma 24

$$\left| \sum_{x=1}^{P} e^{2\pi i f(x)} \right|^{4k_1k_2+2k_2} \qquad (228)$$

$$\leqslant (8\,k_1k_2)^n 2^{4(k_1k_2+k_2+n^2)} P^{4k_1k_2-2k_1+\frac{n(n+1)}{2}+\frac{1}{2}} N_{k_1,s}(P) N_{k_2,n}(P).$$

Now we use the mean value theorem in the form indicated in Theorem 16:

$$N_{k,n}(P) \leqslant (2k)^{2k} (2n)^{n^3} P^{2k-\frac{n(n+1)}{2}+\varepsilon_\tau},$$

where $P \geqslant 1$, $\tau \geqslant 0$, $k = \frac{n(n+1)}{2} + n\tau$ and

$$\varepsilon_\tau = \frac{n(n-1)}{2}\left(1-\frac{1}{n}\right)^{\tau}.$$

Determine τ_1 and τ_2 by the equalities

$$\tau_1 = 2s, \qquad \tau_2 = 1 + \left[2n \log \frac{3\,n}{s}\right].$$

Then, obviously,

$$\varepsilon_{\tau_1} = \frac{s(s-1)}{2}\left(1-\frac{1}{s}\right)^{2s} \leqslant \frac{s(s-1)}{2e^2} < \frac{s^2}{2e^2},$$

$$\varepsilon_{\tau_2} \leqslant \frac{n(n-1)}{2}\left(1-\frac{1}{n}\right)^{2n\left(1+\log \frac{3n}{s}\right)} \leqslant \frac{n(n-1)s^2}{18\,n^2} < \frac{s^2}{18}.$$

Therefore, choosing

$$k_1 = \frac{s(s+1)}{2} + 2s^2 \quad \text{and} \quad k_2 = \frac{n(n+1)}{2} + n + n\left[2\,n \log \frac{3n}{s}\right],$$

we obtain

$$N_{k_1,s}(P) N_{k_2,n}(P) \leqslant (2k_1)^{2k_1} (2k_2)^{2k_2} (2s)^{s^3} (2n)^{n^3} P^{2k_1+2k_2-\frac{n(n+1)}{2}-\frac{s(s+1)}{2}+\varepsilon_{\tau_1}+\varepsilon_{\tau_2}}.$$

Now, observing that $1 \leqslant s \leqslant n + \frac{1}{2}$, $2k_1 + 1 \leqslant 7s^2$ and $\varepsilon_{\tau_1} + \varepsilon_{\tau_2} \leqslant \frac{1}{8}s^2$, from (228) we get

$$\left|\sum_{x=1}^{P} e^{2\pi i f(x)}\right|^{4k_1k_2+2k_2} \leqslant C(s,n)P^{4k_1k_2+2k_2+\frac{1}{2}-\frac{s(s+1)}{2}+\varepsilon_{\tau_1}+\varepsilon_{\tau_2}}$$

$$\leqslant C(s,n)P^{4k_1k_2+2k_2-\frac{3}{8}s^2} \leqslant C(s,n)P^{4k_1k_2+2k_2-\frac{2k_1+1}{19}},$$

where

$$C(s,n) = (2k_1)^{2k_1}(2k_2)^{2k_2}(2s)^{s^3}(2n)^{n^3}(8k_1k_2)^n 2^{4k_1k_2+4k_2+4n^2} < 2^{4k_1k_2+2k_2} n^{40n^3}.$$

Hence, because under $n > 2$

$$k_2 < 2.5\,n^2 \log\frac{3n}{s} \quad \text{and} \quad 2k_1k_2 + k_2 > 10s^2n^2\log\frac{3n}{s},$$

the theorem follows:

$$\left|\sum_{x=1}^{P} e^{2\pi i f(x)}\right| \leqslant 2\,n^{\frac{20n^3}{2k_1k_2+k_2}} P^{1-\frac{1}{38k_2}}$$

$$\leqslant 2\,e^{\frac{2n\log n}{s^2(\log 3n-\log s)}} P^{1-\frac{1}{95n^2(\log 3n-\log s)}}. \tag{229}$$

Note, that the strongest estimates in Theorem 19 are obtained under large values of s. So, for instance, under even n and $s = \frac{n}{2}$ it follows from (229) that

$$\left|\sum_{x=1}^{P} e^{2\pi i f(x)}\right| \leqslant 11\,P^{1-\frac{1}{172n^2}}.$$

Under an arbitrary ε $(0 < \varepsilon \leqslant \frac{1}{2})$ for any $s > \varepsilon n$ we have the estimate

$$\left|\sum_{x=1}^{P} e^{2\pi i f(x)}\right| \leqslant CP^{1-\frac{\gamma}{n^2}}$$

with constants C and γ depending only on ε. Finally, the estimates of the form

$$\left|\sum_{x=1}^{P} e^{2\pi i f(x)}\right| \leqslant CP^{1-\frac{\gamma}{n^2\log\log n}}$$

and

$$\left|\sum_{x=1}^{P} e^{2\pi i f(x)}\right| \leqslant CP^{1-\frac{\gamma}{n^2\log n}},$$

where C and γ are absolute constants, follow from Theorem 19 under $s \geqslant \frac{n}{\log n}$ and $s \geqslant \sqrt{n}$, respectively.

§ 16. Sums arising in zeta-function theory

In investigating a problem on a bound of zeros of the Riemann zeta-function there arises necessity to obtain nontrivial estimates for sums of the form

$$S(t,Q) = \sum_{z=Q+1}^{Q+Q_1} z^{it} \qquad (Q_1 \leqslant Q). \tag{230}$$

The strongest estimates of such sums are obtained with the help of the repeated application of the mean value theorem, which was suggested in [24]–[26] and applied later in [46], [27], [28], [47], and in a series of other papers.

At first we shall prove a lemma similar to Lemma 24.

LEMMA 25. *Under any positive integers P, n, and k for the sum*

$$S = \sum_{x=1}^{P} \sum_{y=1}^{P} e^{2\pi i(\alpha_1 xy + \ldots + \alpha_n x^n y^n)}$$

we have the estimate

$$|S|^{4k^2} \leqslant P^{8k^2-4k} V,$$

where

$$V = \sum_{\lambda_1,\ldots,\lambda_n} N_k^{(P)}(\lambda_1,\ldots,\lambda_n) \left| \sum_{x=1}^{P} e^{2\pi i(\alpha_1\lambda_1 x + \ldots + \alpha_n \lambda_n x^n)} \right|^{2k} \tag{231}$$

and the summation is extended over the region

$$|\lambda_\nu| < kP^\nu, \qquad |\mu_\nu| < kP^\nu \qquad (\nu = 1,2,\ldots,n).$$

Proof. Using the inequality (142), we obtain

$$\begin{aligned} |S|^{2k} &\leqslant \left(\sum_{x=1}^{P} \left| \sum_{y=1}^{P} e^{2\pi i(\alpha_1 xy + \ldots + \alpha_n x^n y^n)} \right| \right)^{2k} \\ &\leqslant P^{2k-1} \sum_{x=1}^{P} \left| \sum_{y=1}^{P} e^{2\pi i(\alpha_1 xy + \ldots + \alpha_n x^n y^n)} \right|^{2k}. \end{aligned} \tag{232}$$

Since by (159)

$$\left| \sum_{y=1}^{P} e^{2\pi i(\alpha_1 xy + \ldots + \alpha_n x^n y^n)} \right|^{2k} = \sum_{\lambda_1,\ldots,\lambda_n} N_k^{(P)}(\lambda_1,\ldots,\lambda_n) e^{2\pi i(\alpha_1\lambda_1 x + \ldots + \alpha_n\lambda_n x^n)},$$

where the summation is extended over the region

$$|\lambda_1| < kP, \ldots, |\lambda_n| < kP^n,$$

then it follows from (232) that

$$|S|^{2k} \leqslant P^{2k-1} \sum_{\lambda_1,\ldots,\lambda_n} N_k^{(P)}(\lambda_1,\ldots,\lambda_n)\left|\sum_{x=1}^{P} e^{2\pi i(\alpha_1\lambda_1 x+\ldots+\alpha_n\lambda_n x^n)}\right|.$$

Raise this inequality to the power $2k$ and use the inequality (141):

$$|S|^{4k^2} \leqslant P^{4k^2-2k}\left(\sum_{\lambda_1,\ldots,\lambda_n} N_k^{(P)}(\lambda_1,\ldots,\lambda_n)\left|\sum_{x=1}^{P} e^{2\pi i(\alpha_1\lambda_1 x+\ldots+\alpha_n\lambda_n x^n)}\right|\right)^{2k}$$

$$\leqslant P^{4k^2-2k}\left(\sum_{\lambda_1,\ldots,\lambda_n} N_k^{(P)}(\lambda_1,\ldots,\lambda_n)\right)^{2k-1}$$

$$\times \sum_{\lambda_1,\ldots,\lambda_n} N_k^{(P)}(\lambda_1,\ldots,\lambda_n)\left|\sum_{x=1}^{P} e^{2\pi i(\alpha_1\lambda_1 x+\ldots+\alpha_n\lambda_n x^n)}\right|^{2k}.$$

Hence, since by (164)

$$\sum_{\lambda_1,\ldots,\lambda_n} N_k^{(P)}(\lambda_1,\ldots,\lambda_n) = P^{2k},$$

we obtain the assertion of the lemma:

$$|S|^{4k^2} \leqslant P^{8k^2-4k} \sum_{\lambda_1,\ldots,\lambda_n} N_k^{(P)}(\lambda_1,\ldots,\lambda_n)\left|\sum_{x=1}^{p} e^{2\pi i(\alpha_1\lambda_1 x+\ldots+\alpha_n\lambda_n x^n)}\right|^{2k}$$

$$= P^{8k^2-4k}V.$$

COROLLARY. *If under* $\nu = 1, 2, \ldots, n$

$$\alpha_\nu = \frac{a_\nu}{q_\nu} + \frac{\theta_\nu}{q_\nu^2}, \qquad (a_\nu, q_\nu) = 1, \quad |\theta_\nu| \leqslant 1,$$

then under any positive integer

$$k \geqslant \frac{n(n+1)}{2}$$

for the sum

$$S = \sum_{x,\,y=1}^{P} e^{2\pi i(\alpha_1 xy+\ldots+\alpha_n x^n y^n)}$$

we have the estimate

$$|S|^{2k^2} \leqslant (2k)^{2n} P^{4k^2-2k+\frac{1}{2k}} N_k(P) \prod_{\nu=1}^{n} \min\left(P^\nu, \sqrt{q_\nu} + \frac{P^\nu}{\sqrt{q_\nu}}\right).$$

Proof. For the quantity V determined by the equality (231) we get

$$V = \sum_{\lambda_1,\ldots,\lambda_n} N_k^{(P)}(\lambda_1,\ldots,\lambda_n)\left|\sum_{x=1}^{P} e^{2\pi i(\alpha_1\lambda_1 x+\ldots+\alpha_n\lambda_n x^n)}\right|^{2k}$$

$$\leqslant N_k(P)\sum_{\lambda_1,\ldots,\lambda_n}\left|\sum_{x=1}^{P} e^{2\pi i(\alpha_1\lambda_1 x+\ldots+\alpha_n\lambda_n x^n)}\right|^{2k}$$

$$= N_k(P)\sum_{\mu_1,\ldots,\mu_n} N_k^{(P)}(\mu_1,\ldots,\mu_n)\sum_{\lambda_1,\ldots,\lambda_n} e^{2\pi i(\alpha_1\lambda_1\mu_1+\ldots+\alpha_n\lambda_n\mu_n)}.$$

Hence it follows by Lemma 1 that

$$V \leqslant N_k(P)\sum_{\mu_1,\ldots,\mu_n} N_k^{(P)}(\mu_1,\ldots,\mu_n)\left|\sum_{|\lambda_1|<kP} e^{2\pi i\,\alpha_1\mu_1\lambda_1}\right|\ldots\left|\sum_{|\lambda_n|<kP^n} e^{2\pi i\,\alpha_n\mu_n\lambda_n}\right|$$

$$\leqslant \left[N_k(P)\right]^2\sum_{\mu_1,\ldots,\mu_n}\min\left(2kP,\frac{1}{2\|\alpha_1\mu_1\|}\right)\ldots\min\left(2kP^n,\frac{1}{2\|\alpha_n\mu_n\|}\right)$$

$$\leqslant (2k)^n\left[N_k(P)\right]^2\prod_{\nu=1}^{n}\sum_{|\mu_\nu|<kP^\nu}\min\left(P^\nu,\frac{1}{\|\alpha_\nu\mu_\nu\|}\right). \tag{233}$$

According to the note of Lemma 14 under any ε from the interval $(0,1]$ we have

$$\sum_{|\mu_\nu|<kP^\nu}\min\left(P^\nu,\frac{1}{\|\alpha_\nu\mu_\nu\|}\right)$$

$$\leqslant \frac{8}{\varepsilon}\left(1+\frac{kP^\nu}{q_\nu}\right)(P^\nu+q_\nu)P^{\nu\varepsilon} \leqslant \frac{8k}{\varepsilon}\,P^{\nu\varepsilon}\left(\sqrt{q_\nu}+\frac{P^\nu}{\sqrt{q_\nu}}\right)^2.$$

Since the trivial estimate

$$\sum_{|\mu_\nu|<kP^\nu}\min\left(P^\nu,\frac{1}{\|\alpha_\nu\mu_\nu\|}\right) < 2kP^{2\nu},$$

is always valid, then choosing $\varepsilon = \frac{1}{k^2}$ we obtain from (233)

$$V \leqslant (2k)^n\left[N_k(P)\right]^2\prod_{\nu=1}^{n}\min\left[2kP^{2\nu},\frac{8k}{\varepsilon}\,P^{\nu\varepsilon}\left(\sqrt{q_\nu}+\frac{P^\nu}{\sqrt{q_\nu}}\right)^2\right]$$

$$\leqslant (2k)^{4n}\left[N_k(P)\right]^2P^{\frac{1}{k}}\prod_{\nu=1}^{n}\min\left[P^{2\nu},\left(\sqrt{q_\nu}+\frac{P^\nu}{\sqrt{q_\nu}}\right)^2\right].$$

Now, using Lemma 25, we arrive at the corollary assertion:

$$|S|^{2k^2} \leqslant \left(P^{8k^2-4k}V\right)^{\frac{1}{2}}$$
$$\leqslant (2k)^{2n} N_k(P) P^{4k^2-2k+\frac{1}{2k}} \prod_{\nu=1}^{n} \min\left(P^\nu, \sqrt{q_\nu} + \frac{P^\nu}{\sqrt{q_\nu}}\right).$$

To estimate the sum (230), we need two simple lemmas more. Denote by $\sum_{x\in M}$ the sum extended over integers x belonging to a certain finite set M.

LEMMA 26. *Let functions $f_1(x)$ and $f_2(x)$ be defined under $x \in M$. Then*

$$\sum_{x\in M} e^{2\pi i(f_1(x)+f_2(x))} = \sum_{x\in M} e^{2\pi i f_1(x)} + 2\pi\theta \sum_{x\in M} |f_2(x)|,$$

where $|\theta| \leqslant 1$.

Proof. Since, obviously,

$$\left|e^{2\pi i f_2(x)} - 1\right| = 2\left|\sin \pi f_2(x)\right| \leqslant 2\pi\left|f_2(x)\right|,$$

then

$$\left|\sum_{x\in M} e^{2\pi i(f_1(x)+f_2(x))} - \sum_{x\in M} e^{2\pi i f_1(x)}\right| = \left|\sum_{x\in M} e^{2\pi i f_2(x)}\left(e^{2\pi i f_1(x)} - 1\right)\right|$$
$$\leqslant \sum_{x\in M} \left|e^{2\pi i f_2(x)} - 1\right| \leqslant 2\pi \sum_{x\in M} \left|f_2(x)\right|$$

and, therefore,

$$\sum_{x\in M} e^{2\pi i(f_1(x)+f_2(x))} = \sum_{x\in M} e^{2\pi i f_1(x)} + 2\pi\theta \sum_{x\in M} |f_2(x)|,$$

where $|\theta| \leqslant 1$.

LEMMA 27. *Let n, P, Q, Q_1 be positive integers, $Q_1 \leqslant Q$, and $P < \sqrt{Q}$. Then under any $t > 0$ we have the estimate*

$$\left|\sum_{z=Q+1}^{Q+Q_1} z^{it}\right| \leqslant \frac{Q}{P^2}\left|\sum_{x,\,y=1}^{P} e^{2\pi i(\alpha_1 xy+\ldots+\alpha_n x^n y^n)}\right| + 2P^2 + Qt\left(\frac{P^2}{Q}\right)^{n+1},$$

where

$$\alpha_\nu = \frac{(-1)^{\nu-1}t}{2\pi\,\nu q^\nu} \qquad (\nu = 1, 2, \ldots, n)$$

and q is a certain integer from the interval $(Q,\ 2Q]$.

Proof. Let function $f(z)$ be determined by the equality

$$f(z) = \frac{t}{2\pi} \log (Q + z).$$

Then using Lemma 19 we obtain

$$\left| \sum_{z=Q+1}^{Q+Q_1} z^{it} \right| = \left| \sum_{z=1}^{Q_1} e^{2\pi i f(z)} \right| \leqslant \frac{1}{P^2} \sum_{z=1}^{Q_1} \left| \sum_{x,y=1}^{P} e^{2\pi i f(z+xy)} \right| + 2P^2$$

$$\leqslant \frac{Q}{P^2} \left| \sum_{x,y=1}^{P} e^{2\pi i f(z_0+xy)} \right| + 2P^2, \qquad (234)$$

where

$$\left| \sum_{x,\, y=1}^{P} e^{2\pi i f(z_0+xy)} \right| = \max_{1 \leqslant z \leqslant Q_1} \left| \sum_{x,y=1}^{P} e^{2\pi i f(z+xy)} \right|.$$

Denote by q the sum $Q + z_0$. Then, obviously, $Q < q \leqslant 2Q$ and

$$\begin{aligned} f(z_0 + xy) &= \frac{t}{2\pi} \log q + \frac{t}{2\pi} \log \left(1 + \frac{xy}{q}\right) \\ &= \frac{t}{2\pi} \log q + \frac{t}{2\pi} \left(\frac{xy}{q} - \frac{x^2y^2}{2q^2} + \ldots \pm \frac{x^n y^n}{nq^n} \right) + \theta(x,\, y) \frac{t}{2\pi} \left(\frac{P^2}{Q} \right)^{n+1} \\ &= \frac{t}{2\pi} \log q + \alpha_1 xy + \ldots + \alpha_n x^n y^n + \theta(x,\, y) \frac{t}{2\pi} \left(\frac{P^2}{Q} \right)^{n+1}, \end{aligned}$$

where $0 \leqslant \theta(x, y) \leqslant 1$. Hence, by Lemma 26 it follows that

$$\left| \sum_{x,y=1}^{P} e^{2\pi i f(z_0+xy)} \right| = \left| \sum_{x,y=1}^{P} e^{2\pi i (\alpha_1 xy + \ldots + \alpha_n x^n y^n)} + \theta(x, y) \frac{t}{2\pi} \left(\frac{P^2}{Q} \right)^{n+1} \right|$$

$$\leqslant \left| \sum_{x,y=1}^{P} e^{2\pi i (\alpha_1 xy + \ldots + \alpha_n x^n y^n)} \right| + tP^2 \left(\frac{P^2}{Q} \right)^{n+1}.$$

We obtain the lemma assertion by substituting this estimate into (234).

THEOREM 20. *Let n, Q, Q_1 be positive integers, even $n \geqslant 12$ and $Q_1 \leqslant Q$. If t belongs to the interval $Q^{\frac{n-1}{3}} \leqslant t < Q^{\frac{n}{3}}$, then*

$$\left| \sum_{z=Q+1}^{Q+Q_1} z^{it} \right| \leqslant 3\, Q^{1 - \frac{1}{2800\, n^2}}.$$

Proof. Choose $P = [Q^{\frac{1}{3}}]$. Then

$$2P^2 + Qt\left(\frac{P^2}{Q}\right)^{n+1} \leqslant 2Q^{\frac{2}{3}} + tQ^{1-\frac{n+1}{3}} < 3Q^{\frac{2}{3}}$$

and the assertion of Lemma 26 may be written in the form

$$\left|\sum_{z=Q+1}^{Q+Q_1} z^{it}\right| \leqslant \frac{Q}{P^2}|S| + 3Q^{\frac{2}{3}}, \tag{235}$$

where

$$S = \sum_{x,y=1}^{P} e^{2\pi i\,(\alpha_1 xy + \ldots + \alpha_n x^n y^n)},$$

$$\alpha_\nu = \frac{(-1)^{\nu-1}}{2\pi\,\nu q^\nu t^{-1}} \qquad (\nu = 1, 2, \ldots, n)$$

and q is a certain integer from the interval $(Q,\ 2Q]$.

Write the quantities α_ν in the form

$$\alpha_\nu = \frac{a_\nu}{q_\nu} + \frac{\theta_\nu}{q_\nu^2}, \qquad (a_\nu, q_\nu) = 1, \quad |\theta_\nu| \leqslant 1, \tag{236}$$

and use the corollary of Lemma 25:

$$|S|^{2k^2} \leqslant (2k)^{2n} P^{4k^3 - 2k + \frac{1}{2k}} N_k(P) \prod_{\nu=1}^{n} \min\left(P^\nu, \sqrt{q_\nu} + \frac{P^\nu}{\sqrt{q_\nu}}\right). \tag{237}$$

Applying the trivial estimate under $\nu < \frac{n}{2}$ and $\nu = n$, we get

$$\prod_{\nu=1}^{n} \min\left(P^\nu, \sqrt{q_\nu} + \frac{P^\nu}{\sqrt{q_\nu}}\right) = P^{\frac{n(n+1)}{2}} \prod_{\nu=1}^{n} \min\left(1, \frac{1}{\sqrt{q_\nu}} + \frac{\sqrt{q_\nu}}{P^\nu}\right)$$
$$\leqslant P^{\frac{n(n+1)}{2}} \prod_{\frac{n}{2} \leqslant \nu < n} \left(\frac{1}{\sqrt{q_\nu}} + \frac{\sqrt{q_\nu}}{P^\nu}\right). \tag{238}$$

Let $\alpha = \frac{1}{\beta}$ and $\beta \geqslant 1$. Then

$$\alpha = \frac{1}{[\beta]} - \frac{\{\beta\}}{\beta[\beta]} = \frac{1}{[\beta]} + \frac{\theta}{[\beta]^2}, \qquad |\theta| < 1. \tag{239}$$

Since under $\nu \geqslant \frac{n}{2}$

$$2\pi\,\nu q^\nu t^{-1} \geqslant 2\pi\,\nu Q^{\nu - \frac{n}{3}} > 1,$$

then by (239)

$$\alpha_\nu = \frac{(-1)^{\nu-1}}{2\pi\,\nu q^\nu t^{-1}} = \frac{(-1)^{\nu-1}}{[2\pi\,\nu q^\nu t^{-1}]} + \frac{\theta_\nu}{[2\pi\,\nu q^\nu t^{-1}]^2}, \qquad |\theta_\nu| < 1,$$

and, therefore, under $\nu \geqslant \frac{n}{2}$ we may choose $q_\nu = [2\pi\,\nu q^\nu t^{-1}]$ in the equalities (236). But then for every ν from the interval $\frac{n}{2} \leqslant \nu < n$ the following estimates are satisfied

$$Q^{\nu-\frac{n}{3}} < q_\nu < 2\pi\,\nu 2^\nu Q^{\nu-\frac{n-1}{3}},$$

$$\frac{1}{\sqrt{q_\nu}} + \frac{\sqrt{q_\nu}}{P^\nu} \leqslant Q^{-\frac{\nu}{2}+\frac{n}{6}} + \sqrt{2\pi\nu}\,2^{\frac{\nu}{2}} Q^{\frac{\nu}{2}-\frac{n-1}{6}}\left(\frac{1}{2}Q^{\frac{1}{3}}\right)^{-\nu} < 2^{2n} Q^{\frac{\nu-n+1}{6}},$$

$$\prod_{\frac{n}{2}\leqslant\nu<n}\left(\frac{1}{\sqrt{q_\nu}} + \frac{\sqrt{q_\nu}}{P^\nu}\right) < 2^{2n^2}\prod_{\frac{n}{2}\leqslant\nu<n} Q^{\frac{\nu-n+1}{6}} = 2^{2n^2} Q^{-\frac{n^2-2n}{48}}.$$

Now we get from (238) and (237)

$$\prod_{\nu=1}^{n}\min\left(P^\nu, \sqrt{q_\nu} + \frac{P^\nu}{\sqrt{q_\nu}}\right) \leqslant 2^{2n^2} P^{\frac{n(n+1)}{2}} Q^{-\frac{n^2-2n}{48}},$$

$$|S|^{2k^2} \leqslant (2k)^{2n} 2^{2n^2} P^{4k^2-2k+\frac{1}{2k}+\frac{n(n+1)}{2}} N_k(P) Q^{-\frac{n^2-2n}{48}}. \tag{240}$$

Choose $\tau = 3n$ and $k = \frac{n(n+1)}{2} + n\tau$. Then by Theorem 16

$$N_k(P) \leqslant (2k)^{2k}(2n)^{n^3} P^{2k-\frac{n(n+1)}{2}+\varepsilon_\tau} \leqslant (2n)^{3n^3} P^{2k-\frac{n(n+1)}{2}+\varepsilon_\tau},$$

where

$$\varepsilon_\tau = \frac{n(n-1)}{2}\left(1-\frac{1}{n}\right)^{3n} < \frac{n^2-n}{40}. \tag{241}$$

Therefore, we obtain from (240)

$$|S|^{2k^2} \leqslant (2k)^{2n} 2^{2n^2} (2n)^{3n^3} P^{4k^2+\frac{1}{2k}+\frac{n^2-n}{40}} Q^{-\frac{n^2-2n}{48}}$$

$$\leqslant (2n)^{4n^3} P^{4k^2} Q^{-\frac{n^2-3n}{80}}.$$

Since $n \geqslant 12$, then $n^2 - 3n \geqslant \frac{3}{4}n^2$ and $9n^4 < k^2 < 13n^4$. But then

$$|S| \leqslant (2n)^{\frac{2n^3}{k^2}} P^2 Q^{-\frac{3n^2}{640k^2}} < 2P^2\,Q^{-\frac{1}{2800\,n^2}}.$$

Substituting this estimate into (235), we arrive at the theorem assertion:

$$\left|\sum_{x=Q+1}^{Q+Q_1} z^{it}\right| \leqslant 2\,Q^{1-\frac{1}{2800\,n^2}} + 3\,Q^{\frac{2}{3}} < 3\,Q^{1-\frac{1}{2800\,n^3}}.$$

§ 17. Incomplete rational sums

Let $n > 2$ and $f(x) = a_1x + \ldots + a_nx^n$ be a polynomial with integral coefficients. Consider the rational exponential sum

$$S(P) = \sum_{x=1}^{P} e^{2\pi i \frac{f(x)}{q}}.$$

If $(a_n, q) = 1$ and $q = P^r$, then under $1 \leqslant r \leqslant n-1$ the estimate

$$|S(P)| \leqslant e^{3n} P^{1-\frac{1}{9n^2 \log n}}$$

follows from Theorem 17. Using the repeated application of the mean value theorem, this result can be slightly strengthened. So it follows from Theorem 19 that for a certain interval of values r we have the estimate

$$|S(P)| \leqslant CP^{1-\frac{\gamma}{n^2}}, \tag{242}$$

where C and γ are absolute constants.

Under an arbitrary positive integer q the estimate (242) is the best among known ones and no approaches to the problem of its essential improvement are seen for the present. But this estimate can be strengthened under a special choice of the denominator q. Let us show how it can be done under q being equal to a power of a prime.

LEMMA 28. *Let α, n, P be positive integers, $p > n^2$ be a prime, $F(x) = b_0 + b_1x + \ldots + b_nx^n$, and $T_\alpha[F, P]$ be the number of solutions of the congruence*

$$F(x) \equiv 0 \pmod{p^\alpha}, \qquad 0 \leqslant x < P. \tag{243}$$

If $(b_0, \ldots, b_n, p) = 1$ and $P \geqslant p^{\frac{\alpha}{n}}$, then the estimate

$$T_\alpha[F, P] \leqslant 2nPp^{-\frac{\alpha}{2n}}$$

holds.

Proof. At first we shall show that under $P = ap^s$ with $1 \leqslant a < p$ and $s \geqslant 0$ we have the estimate

$$T_\alpha[F, ap^s] \leqslant nap^{s-\frac{\alpha}{2n}}. \tag{244}$$

Under $s = 0$ this estimate is trivial, because

$$p^{\frac{\alpha}{n}} \leqslant P = a < p,$$

and, therefore, $nap^{-\frac{\alpha}{2n}} > n$, but the congruence

$$F(x) \equiv 0 \pmod{p^\alpha}, \qquad 0 \leqslant x < a,$$

has at most n solutions.

Let $s \geqslant 1$. Write the congruence (243) in the form

$$F(y+px) \equiv 0 \pmod{p^\alpha}, \qquad 0 \leqslant y < p, \quad 0 \leqslant x < ap^{s-1}, \tag{245}$$

and consider the case $1 \leqslant \alpha \leqslant n$. Passing to the congruence to modulus p, we get

$$F(y) \equiv 0 \pmod{p}, \qquad 0 \leqslant y < p.$$

Hence it is seen that y in (245) may attain at most n distinct values and, therefore,

$$T_\alpha[F, ap^s] \leqslant nap^{s-1} \leqslant nap^{s-\frac{\alpha}{n}} < nap^{s-\frac{\alpha}{2n}}.$$

Now we shall apply the induction. Let us assume that the estimate (244) holds under a certain $\alpha \geqslant n$ and all smaller values α. We should show that the estimation is fulfilled under $\alpha+1$ too.

Indeed, denote by y_1 that solution of the congruence (245), for which the congruence

$$F(y+px) \equiv 0 \pmod{p^{\alpha+1}}, \qquad 0 \leqslant x < ap^{s-1},$$

has the most number of solutions, and determine the polynomial $F_1(x)$ by the equality

$$F_1(x) = p^{-\alpha_1} F(y_1+px),$$

where p^{α_1} is the largest power of p dividing all coefficients of the polynomial $F(y_1+px)$. Note that $\alpha_1 \leqslant n$, because otherwise from the equality

$$F(y_1+px) = F(y_1) + F'(y_1)px + \ldots + \frac{1}{n!}F^{(n)}(y_1)p^n x^n$$

it would follow that $F(y_1) \equiv \ldots \equiv F^{(n)}(y_1) \equiv 0 \pmod{p}$ and $b_0 \equiv \ldots \equiv b_n \equiv 0 \pmod{P}$, which contradicts the hypothesis of the lemma. Reducing the congruence

$$F(y_1+px) \equiv 0 \pmod{p^{\alpha+1}}, \qquad 0 \leqslant x < ap^{s-1},$$

by p^{α_1}, we obtain

$$F_1(x) \equiv 0 \pmod{p^{\alpha+1-\alpha_1}}, \qquad 0 \leqslant x < ap^{s-1}.$$

Since $\alpha_1 \leqslant n$, the number of solutions of this congruence does not exceed the number of solutions of the congruence

$$F_1(x) \equiv 0 \pmod{p^{\alpha+1-n}}, \qquad 0 \leqslant x < ap^{s-1}.$$

Therefore

$$T_{\alpha+1}[F, ap^s] \leqslant nT_{\alpha+1-\alpha_1}\left[F_1, ap^{s-1}\right] \leqslant nT_{\alpha+1-n}\left[F_1, ap^{s-1}\right]. \tag{246}$$

Using $ap^s \geqslant p^{\frac{\alpha+1}{n}}$, we get

$$ap^{s-1} \geqslant p^{\frac{\alpha+1}{n}-1} = p^{\frac{\alpha+1-n}{n}}.$$

But then by the induction hypothesis

$$T_{\alpha+1-n}[F_1, ap^{s-1}] \leqslant nap^{s-1-\frac{\alpha+1-n}{2n}} = nap^{s-\frac{1}{2}-\frac{\alpha+1}{2n}}.$$

Substituting this estimate into (246) and using the condition $n < \sqrt{p}$, we obtain

$$T_{\alpha+1}[F_1, ap^s] \leqslant n^2 ap^{s-\frac{1}{2}-\frac{\alpha+1}{2n}} < nap^{s-\frac{\alpha+1}{2n}}.$$

The proof of the estimate (244) is completed.

The lemma assertion for an arbitrary P follows immediately from the estimate (244). Indeed, determine integers s and a with the help of the conditions

$$p^s \leqslant P < p^{s+1}, \qquad (a-1)p^s \leqslant P < ap^s.$$

Here $s \geqslant 0$ and $1 < a < p$. Since

$$ap^s > P \geqslant p^{\frac{\alpha}{n}},$$

then by (244)

$$T_\alpha[F, P] \leqslant T_\alpha[F, ap^s] \leqslant na\,p^{s-\frac{\alpha}{2n}} = n\left[(a-1)p^s + p^s\right]p^{-\frac{\alpha}{2n}} \leqslant 2nP\,p^{-\frac{\alpha}{2n}}.$$

THEOREM 21. *Let r and α be positive integers, $f(x) = a_1x + \ldots + a_nx^n$, $n \geqslant 35$, $\alpha \geqslant 4n^2$, $p > 4n^2$ be prime, $q = p^\alpha$, $P^r \leqslant q < P^{r+1}$, and $T_\nu(P)$ be the number of solutions of the congruence*

$$f^{(\nu)}(x) \equiv 0 \quad \left(\operatorname{mod} p^{\left[\frac{\alpha}{4}\right]}\right), \qquad 1 \leqslant x \leqslant P.$$

Then under any r from the interval $2 < r \leqslant \frac{n}{5}$ we have the estimate

$$\left|\sum_{x=1}^{P} e^{2\pi i \frac{f(x)}{q}}\right| \leqslant 3P^{1-\frac{\gamma}{r^2}} + nT(P),$$

where γ is an absolute constant and

$$T(P) = \max_{2r+3<\nu\leqslant 3r+3} T_\nu(P).$$

Proof. Determine an integer s with the help of the inequality

$$4(r+1)s \leqslant \alpha < 4(r+1)(s+1). \tag{247}$$

It is easy to verify that the following estimates hold:

$$s \geqslant r+1, \qquad (4r+8) > \alpha, \qquad P > p^{4s}.$$

In fact, if $s \leqslant r$, then we arrive at a contradiction:

$$4n^2 \leqslant \alpha < 4(r+1)(s+1) \leqslant 4(r+1)^2 < 4n^2,$$

and, therefore, $s \geqslant r+1$. Further, it is obvious

$$s(4r+8) \geqslant 4rs + 4s + 4(r+1) = 4(r+1)(s+1) > \alpha.$$

Finally, $P > p^{\frac{\alpha}{r+1}} \geqslant p^{4s}$.

In Lemma 19 we choose $P_1 = a = p^s$. Then we obtain

$$\left| \sum_{x=1}^{P} e^{2\pi i \frac{f(x)}{q}} \right| \leqslant \frac{1}{p^{2s}} \sum_{x=1}^{P} \left| \sum_{y,z=1}^{p^s} e^{2\pi i \frac{f(x+p^s yz)}{p^\alpha}} \right| + 2p^{3s}.$$

Denote by M a set of those x from the interval $1 \leqslant x \leqslant P$, which do not satisfy a single congruence of the form

$$f^{(\nu)}(x) \equiv 0 \quad \left(\bmod\, p^{\left[\frac{\alpha}{4}\right]} \right), \qquad 2r+3 < \nu \leqslant 3r+3.$$

For the remaining $x \in [1, P]$ at least one of the congruences is satisfied and, therefore, the number of such x does not exceed $rT(P)$. Hence, observing that $p^{3s} < P^{\frac{3}{4}}$, we get

$$\left| \sum_{x=1}^{P} e^{2\pi i \frac{f(x)}{q}} \right| \leqslant \frac{1}{p^{2s}} \sum_{x \in M} |S_x| + rT(P) + 2P^{\frac{3}{4}}, \tag{248}$$

where

$$S_x = \sum_{y,z=1}^{p^s} e^{2\pi i \frac{f(x+p^s yz)}{p^\alpha}}.$$

Since by (247)

$$f(x+p^s yz) \equiv f(x) + f'(x)p^s yz + \ldots$$
$$+ \frac{1}{(4r+7)!} f^{(4r+7)}(x)(p^s yz)^{4r+7} \pmod{p^\alpha},$$

then setting $n_1 = 4r + 7$ we obtain

$$|S_x| = \left| \sum_{y,z=1}^{p^s} e^{2\pi i \left(\frac{f'(x)}{p^{\alpha-s}} yz + \ldots + \frac{f^{(n_1)}(x)}{n_1!\, p^{\alpha - n_1 s}} y^{n_1} z^{n_1} \right)} \right|$$

$$= \left| \sum_{y,z=1}^{p^s} e^{2\pi i \left(\frac{b_1}{q_1} yz + \ldots + \frac{b_{n_1}}{q_{n_1}} y^{n_1} z^{n_1} \right)} \right|,$$

where b_ν and q_ν are determined by the equalities

$$\frac{1}{\nu!} \frac{f^{(\nu)}(x)}{p^{\alpha - s\nu}} = \frac{b_\nu}{q_\nu}, \qquad (b_\nu, q_\nu) = 1 \quad (\nu = 1, 2, \ldots, n_1).$$

Let as show that for the quantities q_ν the estimates

$$p^{s(3r+3-\nu)} < q_\nu < p^{s\nu} \qquad (2r + 3 < \nu \leqslant 3r + 3) \tag{249}$$

are valid.

Indeed, since in the sum S_x the quantity x belongs to the set M, for every ν from the interval $2r + 3 < \nu \leqslant 3r + 3$ the greatest common divisor of the numbers $f^\nu(x)$ and p^α is less than $p^{\left[\frac{\alpha}{4}\right]}$ and, therefore,

$$p^{\frac{3\alpha}{4} - s\nu} < q_\nu \leqslant p^{\alpha - s\nu}.$$

Hence, observing that $\frac{3}{4}\alpha - s\nu \geqslant s\,(3r + 3 - \nu)$ and

$$\alpha - s\nu = s\nu + \alpha - 2s\nu \leqslant s\nu + s\,(4r + 8) - 2s\,(2r + 4) = s\nu,$$

we obtain the estimate (249).

Now we shall estimate the sum S_x. According to the corollary of Lemma 25

$$|S_x|^{2k^2} \leqslant (2k)^{2n_1} p^{s\left(4k^2 - 2k + \frac{1}{2k}\right)} N_{k,n_1}(p^s) \prod_{\nu=1}^{n_1} \min\left(p^{s\nu}, \sqrt{q_\nu} + \frac{p^{s\nu}}{\sqrt{q_\nu}} \right). \tag{250}$$

Since by (249) under $2r + 3 < \nu \leqslant 3r + 3$ we have

$$\min\left(p^{s\nu}, \sqrt{q_\nu} + \frac{p^{s\nu}}{\sqrt{q_\nu}} \right) \leqslant 2\, \frac{p^{s\nu}}{\sqrt{q_\nu}},$$

then applying the trivial estimation under remaining ν, we get

$$\prod_{\nu=1}^{n_1} \min\left(p^{s\nu}, \sqrt{q_\nu} + \frac{p^{s\nu}}{\sqrt{q_\nu}} \right) \leqslant 2^r p^{\frac{s\, n_1 (n_1 + 1)}{2}} \prod_{\nu = 2r+4}^{3r+3} q_\nu^{-\frac{1}{2}}.$$

By (249) $q_\nu > p^{s(3r+3-\nu)}$ and, therefore,

$$\prod_{\nu=2r+4}^{3r+3} q_\nu^{-\frac{1}{2}} < \prod_{\nu=2r+4}^{3r+3} p^{-\frac{s(3r+3-\nu)}{2}} = p^{-\frac{sr(r-1)}{4}},$$

$$\prod_{\nu=1}^{n_1} \min\left(p^{s\nu}, \sqrt{q_\nu} + \frac{p^{s\nu}}{\sqrt{q_\nu}}\right) \leqslant 2^r p^{\frac{sn_1(n_1+1)}{2} - \frac{sr(r-1)}{4}}.$$

Substituting this estimate into (250), we obtain

$$|S_x|^{2k^2} \leqslant (4k)^{2n_1} p^{s\left(4k^2-2k+\frac{1}{2k}+\frac{n_1(n_1+1)}{2}-\frac{r^2-r}{4}\right)} N_{k,n_1}(p^s).$$

Now we shall use Theorem 16:

$$N_{k,n_1}(p^s) \leqslant (2k)^{2k}(2n_1)^{n_1^3} p^{2sk-\frac{sn_1(n_1+1)}{2}+s\varepsilon_\tau},$$

where under $\tau = 6n_1$ and $r > 2$

$$k = \frac{n_1(n_1+1)}{2} + 6n_1^2 = 2(4r+7)(13r+23) < 270r^2,$$

$$\varepsilon_\tau = \frac{n_1(n_1+1)}{2}\left(1 - \frac{1}{n_1}\right)^{6n_1} < \frac{n_1(n_1-1)}{800} < \frac{1}{20}r^2.$$

Since the estimates

$$\frac{r^2-r}{4} - \frac{r^2}{20} - \frac{1}{2k} > \frac{r^2}{9}, \qquad rs \geqslant \frac{3(r+1)(s+1)}{5},$$

take place, then using the determination of the quantity s, we get

$$|S_x|^{2k^2} \leqslant (4k)^{2n_1}(2k)^{2k}(2n_1)^{n_1^3} p^{4k^2 s - \frac{r^2 s}{9}} < n_1^{2n_1 k} p^{4k^2 s - \frac{r(r+1)(s+1)}{15}},$$

$$|S_x| < 2p^{2s - \frac{(r+1)(s+1)}{225\cdot 10^4 r^3}} < 2p^{2s - \frac{\alpha}{9\cdot 10^6 r^3}}.$$

Hence the theorem follows by (248):

$$\left|\sum_{x=1}^{P} e^{2\pi i\frac{f(x)}{q}}\right| \leqslant \frac{P}{p^{2s}} \max_{1\leqslant x\leqslant P} |S_x| + rT(P) + 2P^{\frac{3}{4}}$$

$$\leqslant 2Pp^{-\frac{\alpha}{9\cdot 10^6 r^3}} + rT(P) + 2P^{\frac{3}{4}} < 3P^{1-\frac{\gamma}{r^2}} + nT(P),$$

where $\gamma = \frac{1}{9\cdot 10^6}$.

Corollary. *Let* $\beta = \left[\frac{\alpha}{5}\right]$. *If for a polynomial* $f(x) = a_1x + \ldots + a_nx^n$ *the hypotheses of Theorem 21 are satisfied and under* $\nu \geqslant 4r$ *at least one of the coefficients* a_ν *is not divisible by* p^β, *then we have the estimate*

$$\left|\sum_{x=1}^{P} e^{2\pi i \frac{f(x)}{p^\alpha}}\right| \leqslant 3n^2 P^{1-\rho},$$

where $\rho = \gamma \min\left(\frac{r}{n}, \frac{1}{r^2}\right)$ *and* $\gamma > 0$ *is an absolute constant.*

Proof. By Theorem 21

$$\left|\sum_{x=1}^{P} e^{2\pi i \frac{f(x)}{p^\alpha}}\right| \leqslant 3P^{1-\frac{\gamma}{r^2}} + nT(P), \tag{251}$$

where

$$T(P) = \max_{2r+3<\nu\leqslant 3r+3} T_\nu(P)$$

and $T_\nu(P)$ is the number of solutions of the congruence

$$f^{(\nu)}(x) \equiv 0 \quad \left(\bmod\, p^{\left[\frac{\alpha}{4}\right]}\right), \qquad 1 \leqslant x \leqslant P. \tag{252}$$

Denote by β_ν the highest power of p dividing every coefficient $a_\nu, \ldots, a_n$ and determine polynomial $F_\nu(x)$ by the equality

$$F_\nu(x) = p^{-\beta_\nu} f^{(\nu)}(x).$$

Since the congruence (252) is equivalent to the congruence

$$F_\nu(x) \equiv 0 \quad \left(\bmod\, p^{\left[\frac{\alpha}{4}\right]-\beta_\nu}\right), \qquad 1 \leqslant x \leqslant P,$$

and at least one of the coefficients of the polynomial $F_\nu(x)$ is not divisible by p, then by Lemma 28

$$T_\nu(P) \leqslant 2nP\, p^{-\frac{\left[\frac{\alpha}{4}\right]-\beta_\nu}{2(n-\nu)}} \leqslant 2nP\, p^{-\frac{\left[\frac{\alpha}{4}\right]-\beta+1}{2n}} \leqslant 2nP\, p^{-\frac{\alpha}{40n}}.$$

But then the estimate

$$T(P) \leqslant 2nP\, p^{-\frac{\alpha}{40n}} \leqslant 2nP^{1-\frac{r}{40n}}$$

holds and the corollary follows from (251).

§ 18. Double exponential sums

In § 14 in estimating Weyl's sums

$$S(P) = \sum_{x=1}^{P} e^{2\pi i\, f(x)} \tag{253}$$

polynomial $f(x) = \alpha_1 x + \ldots + \alpha_n x^n$ was replaced by polynomial $f(x+y)$ depending on two variables and the estimation for the sum (253) was reduced to the estimation of double exponential sum

$$S = \sum_{x=1}^{P} \sum_{y=1}^{P} e^{2\pi\, F(x,y)}$$

with polynomial

$$F(x,y) = \alpha_1(x+y) + \ldots + \alpha_n(x+y)^n.$$

Another important particular case of double exponential sums with polynomial

$$F(x,y) = \alpha_1 xy + \ldots + \alpha_n x^n y^n$$

was considered in § 16.

We shall show that using the repeated application of the mean value theorem it is easy to obtain ([40], Appendix II) estimates for double exponential sums of a general form

$$S(P_1, P_2) = \sum_{x=1}^{P_1} \sum_{y=1}^{P_2} e^{2\pi i\, F(x,y)}, \tag{254}$$

where

$$F(x,y) = \sum_{j=0}^{n_1} \sum_{k=0}^{n_2} \alpha_{jk} x^j y^k. \tag{255}$$

THEOREM 22. *Under any positive integers n_1, n_2, k_1, k_2, P_1, and P_2 for the double exponential sum (254) we have the estimate*

$$\left|S(P_1,P_2)\right|^{4k_1k_2} \leqslant (2k_2)^{n_2} P_1^{4k_1k_2-2k_1} P_2^{4k_1k_2-2k_2} N_{k_1,n_1}(P_1)\, N_{k_2,n_2}(P_2)\sigma, \tag{256}$$

where

$$\sigma = \sum_{\lambda_1,\ldots,\lambda_{n_1}} \min\left(P_2, \frac{1}{\|\beta_1\|}\right) \ldots \min\left(P_2^{n_2}, \frac{1}{\|\beta_{n_2}\|}\right), \tag{257}$$

the summation is extended over the region $|\lambda_j| < k_1 P_1^j$ $(j = 1,2,\ldots,n_1)$ and the quantities β_k are determined by the equalities

$$\beta_k = \sum_{j=1}^{n_1} \alpha_{jk}\lambda_j \qquad (k = 1,2,\ldots,n_2). \tag{258}$$

Proof. Determine quantities $\alpha_j(y)$ by means of the equalities

$$\alpha_j(y) = \sum_{k=0}^{n_2} \alpha_{jk} y^k \qquad (j = 0, 1, \dots, n_1)$$

and write the polynomial (255) in the form

$$F(x, y) = \sum_{j=0}^{n_1} \left(\sum_{k=0}^{n_2} \alpha_{jk} y^k \right) x^j = \alpha_0(y) + \alpha_1(y)x + \dots + \alpha_{n_1}(y)x^{n_1}.$$

Then, obviously,

$$\begin{aligned}\left|S(P_1, P_2)\right|^{2k_1} &\leqslant P_2^{2k_1-1} \sum_{j=1}^{P_2} \left| \sum_{x=1}^{P_1} e^{2\pi i(\alpha_1(y)x + \dots + \alpha_{n_1}(y)x^{n_1})} \right|^{2k_1} \\ &= P_2^{2k_1-1} \sum_{\lambda_1, \dots, \lambda_{n_1}} N_{k_1}^{(P_1)}(\lambda_1, \dots, \lambda_{n_1}) \sum_{y=1}^{P_2} e^{2\pi i(\alpha_1(y)\lambda_1 + \dots + \alpha_{n_1}(y)\lambda_{n_1})}.\end{aligned}$$

Using the equalities (258), we get

$$\begin{aligned}\alpha_1(y)\lambda_1 + \dots + \alpha_{n_1}(y)\lambda_{n_1} &= \sum_{j=1}^{n_1} \left(\sum_{k=0}^{n_2} \alpha_{jk} y^k \right) \lambda_j \\ &= \sum_{k=0}^{n_2} \left(\sum_{j=1}^{n_1} \alpha_{jk} \lambda_j \right) y^k = \beta_0 + \beta_1 y + \dots + \beta_{n_2} y^{n_2}\end{aligned}$$

and, therefore,

$$\left|S(P_1, P_2)\right|^{2k_1} \leqslant P_2^{2k_1-1} \sum_{\lambda_1, \dots, \lambda_{n_1}} N_{k_1}^{(P_1)}(\lambda_1, \dots, \lambda_{n_1}) \left| \sum_{j=1}^{P_2} e^{2\pi i(\beta_1 y + \dots + \beta_{n_2} y^{n_2})} \right|.$$

Hence reasoning as in the proof of Lemma 25, we obtain the inequality similar to the inequalities established earlier in Lemmas 24 and 25

$$\left|S(P_1, P_2)\right|^{4k_1k_2} \leqslant P_1^{4k_1k_2 - 2k_1} P_2^{4k_1k_2 - 2k_2} V, \tag{259}$$

where

$$\begin{aligned}V &= \sum_{\lambda_1, \dots, \lambda_{n_1}} N_{k_1}^{(P_1)}(\lambda_1, \dots, \lambda_{n_1}) \left| \sum_{y=1}^{P_2} e^{2\pi i(\beta_1 y + \dots + \beta_{n_2} y^{n})} \right|^{2k_2} \\ &= \sum_{\lambda_1, \dots, \mu_{n_2}} N_{k_1}^{(P_1)}(\lambda_1, \dots, \lambda_{n_1}) N_{k_2}^{(P_2)}(\mu_1, \dots, \mu_{n_2}) e^{2\pi i(\beta_1 \mu_1 + \dots + \beta_{n_2} \mu_{n_2})}\end{aligned}$$

and the summation is extended over the region

$$|\lambda_j| < k_1 P_1^j, \qquad |\mu_k| < k_2 P_2^k \qquad (1 \leqslant j \leqslant n_1, \quad 1 \leqslant k \leqslant n_2).$$

Determine quantities β'_j with the help of equalities

$$\beta'_j = \sum_{k=1}^{n_2} \alpha_{jk}\mu_k \qquad (j = 1, 2, \ldots, n_1).$$

Since by (258)

$$\begin{aligned} \beta_1\mu_1 + \ldots + \beta_{n_2}\mu_{n_2} &= \sum_{k=1}^{n_2} \left(\sum_{j=1}^{n_1} \alpha_{jk}\lambda_j \right) \mu_k \\ &= \sum_{j=1}^{n_1} \left(\sum_{k=1}^{n_2} \alpha_{jk}\mu_k \right) \lambda_j = \beta'_1\lambda_1 + \ldots + \beta'_{n_1}\lambda_{n_1}, \end{aligned}$$

then by (159)

$$\begin{aligned} &\sum_{\lambda_1,\ldots,\lambda_{n_1}} N_{k_1}^{(P_1)}(\lambda_1, \ldots, \lambda_{n_1}) e^{2\pi i(\beta_1\mu_1 + \ldots + \beta_{n_2}\mu_{n_2})} \\ &\quad = \sum_{\lambda_1,\ldots,\lambda_{n_1}} N_{k_1}^{(P_1)}(\lambda_1, \ldots, \lambda_{n_1}) e^{2\pi i(\beta'_1\lambda_1 + \ldots + \beta'_{n_1}\lambda_{n_1})} \geqslant 0. \end{aligned}$$

Therefore

$$\begin{aligned} V &= \sum_{\mu_1,\ldots,\mu_{n_2}} N_{k_2}^{(P_2)}(\mu_1, \ldots, \mu_{n_2}) \sum_{\lambda_1,\ldots,\lambda_{n_1}} N_{k_1}^{(P_1)}(\lambda_1, \ldots, \lambda_{n_1}) e^{2\pi i(\beta_1\mu_1 + \ldots + \beta_{n_2}\mu_{n_2})} \\ &\leqslant N_{k_2,n_2}(P_2) \sum_{\lambda_1,\ldots,\lambda_{n_1}} N_{k_1}^{(P_1)}(\lambda_1, \ldots, \lambda_{n_1}) \sum_{\mu_1,\ldots,\mu_{n_2}} e^{2\pi i(\beta_1\mu_1 + \ldots + \beta_{n_2}\mu_{n_2})}. \end{aligned}$$

Hence using Lemma 1 we obtain

$$\begin{aligned} V &\leqslant N_{k_2,n_2}(P_2) \sum_{\lambda_1,\ldots,\lambda_{n_1}} N_{k_1}^{(P_1)}(\lambda_1, \ldots, \lambda_{n_1}) \prod_{\nu=1}^{n_2} \min\left(2\,k_2 P_2^{\nu}, \frac{1}{\|\beta_\nu\|} \right) \\ &\leqslant (2k_2)^{n_2} N_{k_1,n_1}(P_1) N_{k_2,n_2}(P_2)\sigma, \end{aligned}$$

where the quantity σ is defined by the equality (257). Substituting this estimate into (259), we get the theorem assertion.

Theorem 23. *Let $n = 1+\max(n_1, n_2)$, $P_1 \geqslant P_2$, and α_{rs} is an arbitrary coefficient of the polynomial $F(x,y)$. If r and s are not equal to zero,*

$$\alpha_{rs} = \frac{a}{q} + \frac{\theta}{q^2}, \qquad (a,q) = 1, \quad |\theta| \leqslant 1,$$

and the quantity q lies in the interval $P_2^{\frac{1}{2}} \leqslant q \leqslant P_1^r P_2^{s-\frac{1}{2}}$, then for the sum $S(P_1, P_2)$ determined by the equality (254) we have the estimate

$$\left|S(P_1,P_2)\right| \leqslant C(n) P_1 P_2^{1-\frac{\gamma}{(n_1 n_2)^2 \log^2 n}},$$

where γ is an absolute constant and $C(n)$ is a constant depending only on n.

Proof. Theorem 22 reduces the estimation of a double exponential sum to the estimation of the quantity

$$\sigma = \sum_{\lambda_1,\ldots,\lambda_{n_1}} \min\left(P_2, \frac{1}{\|\beta_1\|}\right) \ldots \min\left(P_2^{n_2}, \frac{1}{\|\beta_{n_2}\|}\right),$$

where β_k are determined by the equalities (258). Here we estimate all minima except for the s-th trivially. Then we obtain

$$\sigma \leqslant P_2^{\frac{n_2(n_2+1)}{2}} \sum_{\lambda_1,\ldots,\lambda_{n_1}} \min\left(P_2^s, \frac{1}{\|\beta_s\|}\right).$$

By (258) β_s is a linear function of the quantities $\lambda_1, \ldots, \lambda_{n_1}$ and, in particular, a linear function of λ_r:

$$\beta_s = \alpha_{1s}\lambda_1 + \ldots + \alpha_{n_1 s}\lambda_{n_1} = \alpha_{rs}\lambda_r + \beta.$$

Therefore under $\varepsilon = \frac{1}{4n}$, using the theorem conditions and the note of Lemma 14, we get

$$\sum_{|\lambda_r|<k_1 P_1^r} \min\left(P_2^s, \frac{1}{\|\beta_s\|}\right) = \sum_{|x|<k_1 P_1^r} \min\left(P_2^s, \frac{1}{\|\alpha_{rs}x + \beta\|}\right)$$

$$\leqslant \frac{8}{\varepsilon}\left(1 + \frac{k_1 P_1^r}{q}\right)(P_2^s + q) P_2^{\varepsilon s} \leqslant 128\, n k_1 P_1^r P_2^{s-\frac{1}{4}},$$

$$\sigma \leqslant 64n\,(2k_1)^{n_1} P_1^{\frac{n_1(n_1+1)}{2}} P_2^{\frac{n_2(n_2+1)}{2}-\frac{1}{4}}. \tag{260}$$

Set $\tau_1 = [cn_1 \log n_1]$, $\tau_2 = [cn_2 \log n_2]$, $k_1 = \frac{n_1(n_1+1)}{2} + n_1\tau_1$, $k_2 = \frac{n_2(n_2+1)}{2} + n_2\tau_2$, where c is an absolute constant under which by (194) we have the estimates

$$N_{k_1,n_1}(P_1) \leqslant C_1(n_1) P_1^{2k_1 - \frac{n_1(n_1+1)}{2}},$$

$$N_{k_2,n_2}(P_2) \leqslant C_2(n_2) P_2^{2k_2 - \frac{n_2(n_2+1)}{2}}.$$

Then we get the theorem assertion from (256) and (260):

$$\left|S(P_1, P_2)\right|^{4k_1k_2} \leqslant 64\, n(4k_1k_2)^n C_1(n_1)\, C_2(n_2)\, P_1^{4k_1k_2} P_2^{4k_1k_2 - \frac{1}{4}},$$

$$\left|S(P_1, P_2)\right| \leqslant C(n) P_1 P_2^{1 - \frac{\gamma}{(n_1n_2)^2 \log^2 n}},$$

where $\gamma = \frac{1}{16\,c^2}$.

The results obtained for double exponential sums are without difficulty extended to a case of Weyl's sums of an arbitrary multiplicity m:

$$S(P_1, \ldots, P_m) = \sum_{x_1=1}^{P_1} \cdots \sum_{x_m=1}^{P_m} e^{2\pi i\, F(x_1, \ldots, x_m)},$$

where $F(x_1, \ldots, x_m)$ is a polynomial of m variables

$$F(x_1, \ldots, x_m) = \sum_{\nu_1=0}^{n_1} \cdots \sum_{\nu_m=0}^{n_m} \alpha(\nu_1, \ldots, \nu_m) x_1^{\nu_1} \ldots x_m^{\nu_m}.$$

Let $m \geqslant 2$, $P_1 \geqslant \ldots \geqslant P_m$, $n = 1 + \max(n_1, \ldots, n_m)$, and $\alpha = \alpha(r_1, \ldots, r_m)$ is an arbitrary coefficient of the polynomial $F(x_1, \ldots, x_m)$. If the product $r_1 \ldots r_m \neq 0$,

$$\alpha = \frac{a}{q} + \frac{\theta}{q^2}, \qquad (a, q) = 1, \quad |\theta| \leqslant 1,$$

and the quantity q lies in the interval

$$P_m^{\frac{1}{2}} \leqslant q \leqslant P_1^{r_1} \ldots P_{m-1}^{r_{m-1}} P_m^{r_m - \frac{1}{2}},$$

then we have the estimate

$$\left|S(P_1, \ldots, P_m)\right| \leqslant C P_1 \ldots P_{m-1} P_m^{1 - \frac{\gamma}{(n_1 \ldots n_m)^2 \log^m n}},$$

where C and γ are constants depending on n and m, respectively. A proof of this estimate is obtained with the help of the successive m-fold application of the mean value theorem and in its nature is close to the proof of the analogous estimate for double sums, which was presented in Theorem 23.

Another approach to the estimation of multiple sums is based on a multidimensional generalization of the mean value theorem [1]. This approach allowed to strengthen the sum estimates, which follow from Theorem 22. The strongest estimates of multiple exponential sums are obtained in the article [41]. In the last years numerous publications deal with multiple sums, but interesting applications are known only for double sums arising in estimating ordinary Weyl's sums.

CHAPTER III

FRACTIONAL PARTS DISTRIBUTION, NORMAL NUMBERS, AND QUADRATURE FORMULAS

§ 19. Uniform distribution of fractional parts

The notion of uniform distribution in a general form was introduced by H. Weyl [49]. He obtained also fundamental results concerning functions, whose fractional parts are uniformly distributed.

Let a function $f(x)$ be defined for positive integral values of x. Let us consider a sequence of its fractional parts

$$\{f(1)\},\ \{f(2)\},\ \ldots,\ \{f(P)\},\ \ldots \tag{261}$$

Denote by $N_P(\gamma)$ the number of x ($x = 1, 2, \ldots, P$), which satisfy the inequality

$$\{f(x)\} < \gamma,$$

where γ is an arbitrary number from the interval (0,1].

The sequence (261) is called *uniformly distributed*, if

$$\lim_{P\to\infty} \frac{1}{P} N_P(\gamma) = \gamma. \tag{262}$$

If fractional parts of a function $f(x)$ are uniformly distributed, then this function is said to be uniformly distributed too.

Rewrite the relation (262) in the form

$$N_P(\gamma) = \gamma P + o(P). \tag{263}$$

The equality (263) shows that for uniformly distributed functions under an arbitrary $\gamma \in (0, 1]$ the number of these fractional parts among the first P from the sequence (261), which fall on the interval $[0, \gamma)$, is asymptotically proportional to the length of the interval. If $0 < \gamma_1 < \gamma_2 \leqslant 1$, then the number of fractional parts, which fall on the interval $[\gamma_1, \gamma_2)$, is equal to $N_P(\gamma_2) - N_P(\gamma_1)$ and by (263), evidently, is also asymptotically proportional to the length of this interval:

$$N_P(\gamma_2) - N_P(\gamma_1) = (\gamma_2 - \gamma_1)P + o(P).$$

Let us consider, for example, the function $f(x) = \sqrt{x}$ and show that its fractional parts are uniformly distributed. In fact, denote by $T_k(\gamma)$ the number of satisfactions of the inequality $\{\sqrt{x}\} < \gamma$, when x runs through integers from the interval $k^2 \leqslant x < (k+1)^2$. Then under $n = [\sqrt{P}]$ we get

$$n^2 \leqslant P < (n+1)^2, \qquad T_n(\gamma) \leqslant 2n+1 = O(\sqrt{P}),$$

$$\sum_{k=1}^{n-1} T_k(\gamma) \leqslant N_P(\gamma) < \sum_{k=1}^{n} T_k(\gamma),$$

and, therefore,

$$N_P(\gamma) = \sum_{k=1}^{n-1} T_k(\gamma) + O(\sqrt{P}). \tag{264}$$

Since, obviously, $k = [\sqrt{x}\,]$, then under $x = k^2 + y$ we obtain

$$\{\sqrt{x}\} = \sqrt{k^2+y} - k = \frac{y}{k + \sqrt{k^2+y}}, \qquad 0 \leqslant y \leqslant 2\,k,$$

$$\frac{y}{2\,k+1} \leqslant \{\sqrt{x}\} \leqslant \frac{y}{2\,k}.$$

But then

$$2\,k\gamma \leqslant T_k(\gamma) \leqslant (2\,k+1)\gamma,$$

$$n(n-1)\gamma \leqslant \sum_{k=1}^{n-1} T_k(\gamma) \leqslant (n^2-1)\gamma,$$

$$\sum_{k=1}^{n-1} T_k(\gamma) = \gamma P + O(\sqrt{P}),$$

and by (264)

$$N_P(\gamma) = \gamma P + O(\sqrt{P}).$$

Hence it follows by the definition, that fractional parts $\{\sqrt{x}\}$ are uniformly distributed.

In the same way it is easy to investigate other monotonic functions $f(x)$ satisfying the condition

$$\lim_{x\to\infty} \frac{f(x)}{x} = 0.$$

In particular, it can be shown that under $0 < \alpha < 1$ fractional parts of the function x^α are uniformly distributed, but fractional parts of the function $\log^\beta x$ are distributed uniformly or not, depending on whether $\beta > 1$ or $\beta \leqslant 1$.

The investigation of functions growing as polynomials and especially functions growing faster is much more difficult. So, for example, it is not known whether fractional parts of the functions e^x and $\left(\frac{3}{2}\right)^x$ or of the functions $\left(\frac{m}{n}\right)^x$ under $n > 1$,

$m > n$, and coprime m and n are uniformly distributed. At the same time it is easy to present nontrivial examples of exponential functions, whose fractional parts are not uniformly distributed.

Indeed, let λ and θ be roots of the quadratic equation $z^2 + pz + q = 0$, with integral coefficients, such that $\lambda > 1$ and $0 < \theta < 1$ (it is so, for instance, under $p = -3$ and $q = 1$). Since the symmetric function $\lambda^x + \theta^x$ takes on positive integral values under $x = 1, 2, \ldots$, then $\{\lambda^x\} = 1 - \theta^x$. Therefore fractional parts of the function λ^x monotonically grow, approaching unity, and, evidently, are not uniformly distributed.

We will consider a general criterion for uniform distribution, connecting problems of distribution of fractional parts with estimations of exponential sums.

Let $0 < \gamma < 1$ and $0 < \varepsilon < \min(\gamma,\ 1-\gamma)$. Determine functions $\psi_1(x)$ and $\psi_2(x)$ by means of the equalities

$$\psi_1(x) = \begin{cases} \frac{1}{\varepsilon} x & \text{if } 0 \leqslant x < \varepsilon, \\ 1 & \text{if } \varepsilon \leqslant x < \gamma - \varepsilon, \\ \frac{1}{\varepsilon}(\gamma - x) & \text{if } \gamma - \varepsilon \leqslant x < \gamma, \\ 0 & \text{if } \gamma \leqslant x < 1, \end{cases}$$

$$\psi_2(x) = \begin{cases} 1 & \text{if } 0 \leqslant x < \gamma, \\ \frac{1}{\varepsilon}(\gamma + \varepsilon - x) & \text{if } \gamma \leqslant x < \gamma + \varepsilon, \\ 0 & \text{if } \gamma + \varepsilon \leqslant x < 1 - \varepsilon, \\ \frac{1}{\varepsilon}(x + \varepsilon - 1) & \text{if } 1 - \varepsilon \leqslant x < 1. \end{cases} \tag{265}$$

Lemma 29. *Let $\psi(x)$ be the characteristic function of the interval $[0, \gamma)$. Denote by $C_1(m)$ and $C_2(m)$ the Fourier coefficients of the functions $\psi_1(x)$ and $\psi_2(x)$. Then the relations*

$$\begin{gathered} \psi_1(x) \leqslant \psi(x) \leqslant \psi_2(x) \qquad (0 \leqslant x \leqslant 1), \\ C_1(0) = \gamma - \varepsilon, \qquad C_2(0) = \gamma + \varepsilon, \\ \max\left(|C_1(m)|,\ |C_2(m)|\right) \leqslant \min\left(\frac{1}{\pi|m|},\ \frac{1}{\pi^2 m^2 \varepsilon}\right) \qquad (m = \pm1, \pm2, \ldots) \end{gathered} \tag{266}$$

hold.

Proof. Since

$$\psi(x) = \begin{cases} 1 & \text{if } 0 \leqslant x < \gamma, \\ 0 & \text{if } \gamma \leqslant x < 1, \end{cases}$$

the first of the relations (266) follows directly from the determination of the functions $\psi_1(x)$ and $\psi_2(x)$.

Further, obviously,

$$C_1(0) = \int_0^1 \psi_1(x)\,dx = \frac{1}{\varepsilon}\int_0^{\varepsilon} x\,dx + \int_{\varepsilon}^{\gamma-\varepsilon} dx - \frac{1}{\varepsilon}\int_{\gamma-\varepsilon}^{\nu} (\gamma - x)\,dx$$

$$= \frac{\varepsilon}{2} + \gamma - 2\varepsilon + \frac{\varepsilon}{2} = \gamma - \varepsilon.$$

Analogously we get

$$C_2(0) = \int_0^1 \psi_2(x)\,dx = \gamma + \varepsilon,$$

the second assertion of the lemma is proved.

Finally, under $m \neq 0$ we have

$$C_1(m) = \int_0^1 \psi_1(x) e^{-2\pi i\, mx}\,dx = \frac{1}{\varepsilon}\int_0^{\varepsilon} x e^{-2\pi i\, mx}\,dx + \int_{\varepsilon}^{\gamma-s} e^{-2\pi i\, mx}\,dx$$

$$+ \frac{1}{\varepsilon}\int_{\gamma-s}^{\gamma} (\gamma - x) e^{-2\pi i\, mx}\,dx = \frac{1 - e^{-2\pi i\, m(\gamma-\varepsilon)}}{2\pi i\, m\varepsilon}\int_0^{\varepsilon} e^{-2\pi i\, mx}\,dx.$$

Hence, since

$$\left|\int_0^{\varepsilon} e^{-2\pi i\, mx}\,dx\right| \leqslant \min\left(\varepsilon, \frac{1}{\pi|m|}\right),$$

it follows that

$$|C_1(m)| \leqslant \min\left(\frac{1}{\pi|m|}, \frac{1}{\pi^2 m^2 \varepsilon}\right).$$

In the same way we obtain the estimate

$$|C_2(m)| \leqslant \min\left(\frac{1}{\pi|m|}, \frac{1}{\pi^2 m^2 \varepsilon}\right),$$

the lemma is proved completely.

THEOREM 24 (the Weyl criterion). *A necessary and sufficient condition for uniform distribution of a function $f(x)$ is*

$$\lim_{R\to\infty} \frac{1}{P}\sum_{x=1}^{P} e^{2\pi i\, mf(x)} = 0 \tag{267}$$

for any integer $m \neq 0$.

Proof. Let $0 < \gamma < 1$, $\psi(x)$ be the characteristic function of the interval $[0,\gamma)$ and the functions $\psi_1(x)$ and $\psi_2(x)$ be determined by the equalities (265). Then, evidently

$$N_P(\gamma) = \sum_{x=1}^{P} \psi(\{f(x)\}). \tag{268}$$

Since by the lemma

$$\psi_1(\{f(x)\}) \leqslant \psi(\{f(x)\}) \leqslant \psi_2(\{f(x)\}),$$

then, carrying out the summation over x, by (268) we get

$$\sum_{x=1}^{P} \psi_1(\{f(x)\}) - \gamma P \leqslant N_P(\gamma) - \gamma P \leqslant \sum_{x=1}^{P} \psi_2(\{f(x)\}) - \gamma P. \tag{269}$$

Let us suggest that the condition (267) is satisfied. Choose $P > P_0(\varepsilon)$ in such a way that under $n = \left[\frac{1}{\varepsilon}\right] + 1$ the estimate

$$\max_{1 \leqslant m \leqslant n^2} \left| \sum_{x=1}^{P} e^{2\pi i\, m f(x)} \right| \leqslant \frac{\pi}{4(1 + 2\log n)}\, \varepsilon P \tag{270}$$

would be satisfied. Then, using the expansion of the function $\psi_1(\{f(x)\})$ into the Fourier series, we obtain

$$\psi_1(\{f(x)\}) = \gamma - \varepsilon + \sum_{m=-\infty}^{\infty}{}' \, C_1(m) e^{2\pi i\, m f(x)},$$

$$\left| \sum_{x=1}^{P} \psi_1(\{f(x)\}) - \gamma P \right| = \left| -\varepsilon P + \sum_{m=-\infty}^{\infty}{}' \, C_1(m) \sum_{x=1}^{P} e^{2\pi i\, m f(x)} \right|$$

$$\leqslant \varepsilon P + \sum_{m=-n^2}^{n^2}{}' \, |C_1(m)| \left| \sum_{x=1}^{P} e^{2\pi i\, m f(x)} \right|$$

$$+ \sum_{|m| > n^2} |C_1(m)| \left| \sum_{x=1}^{P} e^{2\pi i\, m f(x)} \right|$$

(the sign $'$ in the sum indicates the deletion of the summand with $m = 0$). Hence applying the estimate (270) under $|m| \leqslant n^2$ and the trivial one under $|m| > n^2$, by virtue of the lemma we get

$$\left| \sum_{x=1}^{P} \psi_1(\{f(x)\}) - \gamma P \right| \leqslant \varepsilon P + \frac{\pi}{4(1 + 2\log n)}\, \varepsilon P \sum_{m=-n^2}^{n^2}{}' \, \frac{1}{\pi |m|} + P \sum_{|m| > n^2} \frac{1}{\pi^2 m^2 \varepsilon}$$

$$\leqslant \varepsilon P + \frac{1}{2}\varepsilon P + \frac{2P}{\pi^2 n^2 \varepsilon} < 2\varepsilon P.$$

In the same way the estimate

$$\left|\sum_{x=1}^{P}\psi_2(\{f(x)\}) - \gamma P\right| \leqslant 2\varepsilon P$$

is obtained. But then it follows from (269) that

$$-2\varepsilon P \leqslant N_P(\gamma) - \gamma P \leqslant 2\varepsilon P$$

and, because ε can be as small as we please, we get

$$\lim_{P\to\infty}\frac{1}{P}N_P(\gamma) = \gamma.$$

The sufficiency of the condition (267) is proved.

Now we shall prove the necessity of that condition. Indeed, let the function $f(x)$ be uniformly distributed. Take $m \neq 0$ and choose an integer $q > |m|$. Denote by M_k a set of those x from the interval $[1, P]$, which satisfy

$$\frac{k}{q} \leqslant \{f(x)\} < \frac{k+1}{q}. \tag{271}$$

Denote by T_k the number of satisfactions of this inequality. Then, obviously,

$$\sum_{x=1}^{P} e^{2\pi i\, mf(x)} = \sum_{k=0}^{q-1}\sum_{x\in M_k} e^{2\pi i\, mf(x)}.$$

It follows from (271) that for $x \in M_k$

$$\{f(x)\} = \frac{k}{q} + \frac{\theta_x}{q}, \qquad 0 \leqslant \theta_x < 1.$$

Using Lemma 26, we obtain

$$\sum_{x\in M_k} e^{2\pi i\, mf(x)} = \sum_{x\in M_k} e^{2\pi i\left(\frac{mk}{q}+\frac{m\theta_x}{q}\right)} = \sum_{x\in M_k} e^{2\pi i\,\frac{mk}{q}} + 2\pi\theta(k)\sum_{x\in M_k}\frac{|m|\theta_x}{q}$$
$$= T_k e^{2\pi i\,\frac{mk}{q}} + \frac{2\pi|m|}{q}\theta'(k)T_k,$$

where $|\theta'(k)| \leqslant 1$. But then

$$\left|\sum_{x=1}^{P} e^{2\pi i\, mf(x)}\right| = \left|\sum_{k=0}^{q-1} T_k e^{2\pi i\,\frac{mk}{q}} + \frac{2\pi|m|}{q}\sum_{k=0}^{q-1}\theta'_k T_k\right|$$
$$\leqslant \left|\sum_{k=0}^{q-1} T_k e^{2\pi i\,\frac{mk}{q}}\right| + \frac{2\pi|m|}{q}\sum_{k=0}^{q-1} T_k$$
$$= \left|\sum_{k=0}^{q-1} T_k e^{2\pi i\,\frac{mk}{q}}\right| + \frac{2\pi|m|}{q}P. \tag{272}$$

Since by the hypothesis fractional parts of the function $f(x)$ are uniformly distributed, then

$$T_k = \frac{1}{q} P + o(P).$$

Take an arbitrary $\varepsilon > 0$ and choose $P > P_0(\varepsilon)$ such that under $q = \left[\frac{4\pi|m|}{\varepsilon}\right] + 1$ the estimate

$$\left|T_k - \frac{1}{q} P\right| \leqslant \frac{1}{2q} \varepsilon P$$

will hold. Observing that $1 \leqslant |m| < q$ and therefore

$$\sum_{k=0}^{q-1} e^{2\pi i \frac{mk}{q}} = 0,$$

we obtain

$$\left|\sum_{k=0}^{q-1} T_k e^{2\pi i \frac{mk}{q}}\right| = \left|\frac{1}{q} P \sum_{k=0}^{q-1} e^{2\pi i \frac{mk}{q}} + \sum_{k=0}^{q-1} \left(T_k - \frac{1}{q} P\right) e^{2\pi i \frac{mk}{q}}\right|$$
$$\leqslant \sum_{k=0}^{q-1} \left|T_k - \frac{1}{q} P\right| \leqslant \frac{1}{2} \varepsilon P.$$

But then, because of $q > \frac{4\pi|m|}{\varepsilon}$, we obtain the equality (267) from (272):

$$\left|\sum_{x=1}^{P} e^{2\pi i\, m f(x)}\right| \leqslant \frac{1}{2} \varepsilon P + \frac{2\pi|m|}{q} P \leqslant \varepsilon P,$$

$$\lim \frac{1}{P}\left|\sum_{x=1}^{P} e^{2\pi i\, m f(x)}\right| = 0.$$

The theorem is proved completely.

As an example of an application of the Weyl criterion we shall show that fractional parts of a linear function

$$f(x) = \alpha x + \beta \tag{273}$$

are uniformly distributed under an irrational α and an arbitrary β.

Indeed, let $m \neq 0$ be an integer. Using Lemma 1, we get

$$\left|\sum_{x=1}^{P} e^{2\pi i\, m(\alpha x + \beta)}\right| = \left|\sum_{x=1}^{P} e^{2\pi i\, \alpha m x}\right| \leqslant \min\left(P, \frac{1}{2\|\alpha m\|}\right) \leqslant \frac{1}{2\|\alpha m\|}.$$

Therefore

$$\frac{1}{P}\left|\sum_{x=1}^{P} e^{2\pi i\, m(\alpha x + \beta)}\right| \leqslant \frac{1}{2\|\alpha m\| P},$$

$$\lim_{P\to\infty}\frac{1}{P}\left|\sum_{x=1}^{P}e^{2\pi i\, m(\alpha x+\beta)}\right| = 0,$$

and by virtue of the Weyl criterion fractional parts of the function (273) are uniformly distributed.

Now we shall consider a question concerning the distribution of fractional parts of a polynomial of an arbitrary degree $n \geqslant 1$:

$$f(x) = \alpha_0 + \alpha_1 x + \ldots + \alpha_n x^n.$$

At first we shall prove a sufficient condition of uniform distribution of fractional parts which is due to van der Corput [7].

LEMMA 30. *Let $\underset{h}{\Delta} f(x)$ be a finite difference of a function $f(x)$ with step h:*

$$\underset{h}{\Delta} f(x) = f(x+h) - f(x).$$

Then under any P_1 from the interval $[1, P]$ we have the estimate

$$\left|\sum_{x=1}^{P} e^{2\pi i\, f(x)}\right|^2 \leqslant 2P\left(PP_1^{-1} + 2P_1 + \max_{1\leqslant h<P_1}\left|\sum_{x=1}^{P} e^{2\pi i\, \underset{h}{\Delta} f(x)}\right|\right).$$

Proof. By (196)

$$\left|\sum_{x=1}^{P} e^{2\pi i\, f(x)}\right| \leqslant \frac{1}{P_1}\sum_{x=1}^{P}\left|\sum_{y=0}^{P_1-1} e^{2\pi i\, f(x+y)}\right| + P_1$$

and, therefore,

$$\left|\sum_{x=1}^{P} e^{2\pi i\, f(x)}\right|^2 \leqslant \frac{2P}{P_1^2}\sum_{x=1}^{P}\left|\sum_{y=0}^{P_1-1} e^{2\pi i\, f(x+y)}\right|^2 + 2P_1^2. \tag{274}$$

Since, obviously,

$$\begin{aligned}\sum_{x=1}^{P}\left|\sum_{y=0}^{P_1-1} e^{2\pi i\, f(x+y)}\right|^2 &\leqslant \sum_{y,z=0}^{P_1-1}\left|\sum_{x=1}^{P} e^{2\pi i\,(f(x+y)-f(x+z))}\right| \\ &= PP_1 + 2\sum_{y>z}\left|\sum_{x=1}^{P} e^{2\pi i\,(f(x+y)-f(x+z))}\right|,\end{aligned} \tag{275}$$

then observing that

$$\left|\sum_{x=1}^{P} e^{2\pi i (f(x+y)-f(x+z))}\right| = \left|\sum_{x=z+1}^{z+P} e^{2\pi i (f(x+y-z)-f(x))}\right|$$
$$\leqslant 2z + \left|\sum_{x=1}^{P} e^{2\pi i (f(x+y-z)-f(x))}\right|,$$

we obtain from (275)

$$\sum_{x=1}^{P}\left|\sum_{y=0}^{P_1-1} e^{2\pi i f(x+y)}\right|^2 \leqslant PP_1 + P_1^3 + 2\sum_{y>z}\left|\sum_{x=1}^{P} e^{2\pi i (f(x+y-z)-f(x))}\right|$$
$$\leqslant PP_1 + P_1^3 + P_1^2 \max_{1\leqslant h<P_1}\left|\sum_{x=1}^{P} e^{2\pi i \underset{h}{\Delta} f(x)}\right|.$$

Substituting this estimate into (274), we get the assertion of the lemma:

$$\left|\sum_{x=1}^{P} e^{2\pi i f(x)}\right|^2 \leqslant 2P^2P_1^{-1} + 2PP_1 + 2P \max_{1\leqslant h<P_1}\left|\sum_{x=1}^{P} e^{2\pi i \underset{h}{\Delta} f(x)}\right| + 2P_1^2$$
$$\leqslant 2P\left(PP_1^{-1} + 2P_1 + \max_{1\leqslant h<P_1}\left|\sum_{x=1}^{P} e^{2\pi i \underset{h}{\Delta} f(x)}\right|\right).$$

THEOREM 25. *A sufficient condition of uniform distribution of a function $f(x)$ is uniform distribution of its finite difference $\underset{h}{\Delta} f(x)$ under any integer $h \geqslant 1$.*

Proof. Let according to the hypothesis of the theorem fractional parts $\{\underset{h}{\Delta} f(x)\}$ be uniformly distributed under any positive integer k. Then by the Weyl criterion under every integer $m \neq 0$

$$\lim_{P\to\infty} \frac{1}{P}\sum_{x=1}^{P} e^{2\pi i m \underset{h}{\Delta} f(x)} = 0. \tag{276}$$

Apply the inequality of Lemma 30 to the function $mf(x)$. Then observing that $\underset{h}{\Delta} mf(x) = m \underset{h}{\Delta} f(x)$ we obtain

$$\left|\sum_{x=1}^{P} e^{2\pi i m f(x)}\right|^2 \leqslant 2P\left(PP_1^{-1} + 2P_1 + \max_{1\leqslant h<P_1}\left|\sum_{x=1}^{P} e^{2\pi i m \underset{h}{\Delta} f(x)}\right|\right). \tag{277}$$

Let $0 < \varepsilon < 1$ and $P_1 = \left[\frac{6}{\varepsilon^2}\right] + 1$. If follows from (276) that $P_2 = P_2(\varepsilon, m)$ can be chosen in such a way that under $P \geqslant \max(P_1, P_2)$ the inequality

$$\max_{1\leqslant h<P_1}\left|\sum_{x=1}^{P} e^{2\pi i m \underset{h}{\Delta} f(x)}\right| < \frac{1}{6}\varepsilon^2 P$$

will be satisfied. Choose $P \geqslant \max\left(\frac{12}{\varepsilon^2}P_1, P_2\right)$. Then we get from (277)

$$\left|\sum_{x=1}^{P} e^{2\pi i\, m f(x)}\right|^2 \leqslant 2P\left(\frac{1}{6}\varepsilon^2 P + \frac{1}{6}\varepsilon^2 P + \frac{1}{6}\varepsilon^2 P\right) = \varepsilon^2 P^2,$$

$$\left|\sum_{x=1}^{P} e^{2\pi i\, m f(x)}\right| \leqslant \varepsilon P,$$

and, therefore,

$$\lim_{P\to\infty} \frac{1}{P}\sum_{x=1}^{P} e^{2\pi i\, m f(x)} = 0.$$

The theorem is proved by virtue of the Weyl criterion.

THEOREM 26 (Weyl's theorem). *If a polynomial*

$$f(x) = \alpha_0 + \alpha_1 x + \ldots + \alpha_n x^n \tag{278}$$

has at least one non constant term with an irrational coefficient, then its fractional parts are uniformly distributed.

Proof. We shall start with a case when the coefficient of the highest degree term is irrational. Under $n = 1$ the polynomial (278) is in reality a linear function $\alpha_0 + \alpha_1 x$ with an irrational coefficient α_1. By (273) fractional parts of such linear functions are uniformly distributed. Apply induction. Let $n \geqslant 2$ and the theorem be proved for polynomials of degree $n-1$, having an irrational coefficient of the highest degree term. Choose an arbitrary positive integer h and consider the finite difference

$$\underset{h}{\Delta} f(x) = f(x+h) - f(x) = \alpha_n\left[(x+h)^n - x^n\right] + \ldots + \alpha_1[(x+h) - x].$$

Evidently, $\underset{h}{\Delta} f(x)$ is a polynomial of the $(n-1)$-th degree with an irrational coefficient of the highest degree term. By the induction hypothesis, fractional parts of this polynomial are uniformly distributed. But then by Theorem 25 fractional parts of the initial polynomial are uniformly distributed also. Thus the theorem is proved for polynomials with the leading coefficient being irrational.

Now let $1 \leqslant s < n$ and α_s be the leading among irrational coefficients of the polynomial $f(x)$. Denote by q the common denominator of coefficients $\alpha_{s+1}, \ldots, \alpha_n$ and write the polynomial (278) in the form

$$f(x) = f_1(x) + \frac{\varphi(x)}{q},$$

where $f_1(x) = \alpha_0 + \alpha_1 x + \ldots + \alpha_s x^s$ and $\varphi(x)$ is a polynomial with integral coefficients. Choose an arbitrary integer $m \neq 0$ and determine an integer P_1 from the condition

$$P_1 q \leqslant P < (P_1 + 1)q.$$

Then, setting $x = y + qz$, we obtain

$$\sum_{x=1}^{P_1 q} e^{2\pi i\, m f(x)} = \sum_{y=1}^{q} \sum_{z=0}^{P_1-1} e^{2\pi i\left(m f_1(y+qz) + \frac{m\varphi(y+qz)}{q} \right)}$$
$$= \sum_{y=1}^{q} e^{2\pi i \frac{m\varphi(y)}{q}} \sum_{z=0}^{P_1-1} e^{2\pi i\, m f_1(y+qz)}.$$

Since the coefficient of the highest power z of the polynomial $f_1(y+qz)$ is irrational, fractional parts of $f_1(y+qz)$ are uniformly distributed and by the Weyl criterion

$$\sum_{z=0}^{P_1-1} e^{2\pi i\, m f(y+qz)} = o(P_1).$$

But then

$$\sum_{x=1}^{P} e^{2\pi i\, m f(x)} = \sum_{y=1}^{q} e^{2\pi i \frac{m\varphi(y)}{q}} \sum_{z=0}^{P_1-1} e^{2\pi i\, m f_1(y+qz)} + O(q) = o(P).$$

Hence, applying the Weyl criterion again, we get the theorem assertion.

§ 20. Uniform distribution of functions systems and completely uniform distribution

Let $s \geqslant 1$ be a fixed positive integer, $\gamma_1, \ldots, \gamma_s$ be arbitrary positive numbers not exceeding 1, and $f_1(x), \ldots, f_s(x)$ be functions defined for positive integral values x. Denote by $N_P(\gamma_1, \ldots, \gamma_s)$ the number of satisfactions of the system of inequalities

$$\left. \begin{array}{l} \{f_1(x)\} < \gamma_1 \\ \ldots\ldots\ldots\ldots \\ \{f_s(x)\} < \gamma_s \end{array} \right\}, \qquad x = 1, 2, \ldots, P.$$

A system of functions $f_1(x), \ldots, f_s(x)$ is called *uniformly distributed* in the s-dimensional unit cube, if

$$\lim_{P\to\infty} \frac{1}{P} N_P(\gamma_1, \ldots, \gamma_s) = \gamma_1 \cdots \gamma_s$$

or, what is just the same,

$$N_P(\gamma_1, \ldots, \gamma_s) = \gamma_1 \cdots \gamma_s P + o(P).$$

It is easily seen that under $s = 1$ this definition is identical with the definition of uniform distribution introduced in the preceding section.

Let $m_1, \dots, m_s$ be arbitrary integers not all zero. In the same way, as in the proof of Theorem 24, it can be shown that a necessary and sufficient condition of uniform distribution of a system of functions $f_1(x), \dots, f_s(x)$ is

$$\lim_{P\to\infty} \frac{1}{P} \sum_{x=1}^{P} e^{2\pi i (m_1 f_1(x) + \dots + m_s f_s(x))} = 0. \tag{279}$$

This equality, representing the multidimensional criterion of Weyl, reduces the investigation of the uniformity of distribution of functions system to estimations of corresponding exponential sums.

Using the multidimensional criterion of Weyl, it is easy to show that a necessary and sufficient condition of uniform distribution of a system of functions $f_1(x), \dots, f_s(x)$ is uniform distribution of the function

$$F(x) = m_1 f_1(x) + \dots + m_s f_s(x) \tag{280}$$

for any integers $m_1, \dots, m_s$ not all zero.

Indeed, if fractional parts of the function $F(x)$ are uniformly distributed, then under any integer $m \neq 0$ the equality

$$\sum_{x=1}^{P} e^{2\pi i mF(x)} = o(P)$$

holds. Hence under $m = 1$ it follows that

$$\sum_{x=1}^{P} e^{2\pi i (m_1 f_1(x) + \dots + m_s f_s(x))} = \sum_{x=1}^{P} e^{2\pi i F(x)} = o(P),$$

and by the multidimensional criterion of Weyl the system of functions $f_1(x), \dots, f_s(x)$ is uniformly distributed. On the other hand, if a system of functions $f_1(x), \dots, f_s(x)$ is uniformly distributed in the s-dimensional unit cube, then by (279)

$$\sum_{x=1}^{P} e^{2\pi i (m_1 f_1(x) + \dots + m_s f_s(x))} = o(P). \tag{281}$$

Choose an arbitrary integer $m \neq 0$ and replace m_ν by mm_ν $(\nu = 1, 2, \dots, s)$ in the equality (281). Then we obtain

$$\sum_{x=1}^{P} e^{2\pi i mF(x)} = \sum_{x=1}^{P} e^{2\pi i (mm_1 f_1(x) + \dots + mm_s f_s(x))} = o(P).$$

Hence it follows by the Weyl criterion, that fractional parts of the function $F(x)$ are uniformly distributed. The property (280) is proved completely.

Let us show that a system of linear functions

$$f_1(x) = \alpha_1 x, \ldots, f_s(x) = \alpha_s x \tag{282}$$

is uniformly distributed in the s-dimensional unit cube under certain requirements for the quantities $\alpha_1, \ldots, \alpha_s$.

Indeed, let the numbers $1, \alpha_1, \ldots, \alpha_s$ be linearly independent. Then under any integers $m_1, \ldots, m_s$ not all zero a linearly combination $m_1\alpha_1 + \ldots + m_s\alpha_s$ cannot be equal to an integer. Therefore, by Lemma 1

$$\left|\sum_{x=1}^{P} e^{2\pi i (m_1 f_1(x) + \ldots + m_s f_s(x))}\right| = \left|\sum_{x=1}^{P} e^{2\pi i (m_1\alpha_1 + \ldots + m_s\alpha_s)x}\right| \leqslant \frac{1}{2\|m_1\alpha_1 + \ldots + m_s\alpha_s\|} = o(P)$$

and the system of functions (282) is uniformly distributed by the Weyl criterion.

Now let the numbers $1, \alpha_1, \ldots, \alpha_s$ be not linearly independent. Then there exist integers $m_1, \ldots, m_{s+1}$ not all zero such that $m_1\alpha_1 + \ldots + m_s\alpha_s = m_{s+1}$. Therefore, under $m_1, \ldots, m_s$ satisfying this equality we get

$$\sum_{x=1}^{P} e^{2\pi i (m_1 f_1(x) + \ldots + m_s f_s(x))} = \sum_{x=1}^{P} e^{2\pi i (m_1\alpha_1 + \ldots + m_s\alpha_s)x} = \sum_{x=1}^{P} e^{2\pi i\, m_{s+1} x} = P.$$

But then the condition (279) is not satisfied and the system of functions $\alpha_1 x, \ldots, \alpha_s x$ is not uniformly distributed. Thus the system of linear functions (282) is uniformly distributed in the s-dimensional unit cube if and only if the numbers $1, \alpha_1, \ldots, \alpha_s$ are linearly independent.

THEOREM 27. *Let $f(x) = \alpha_0 + \alpha_1 x + \ldots + \alpha_n x^n$ be a polynomial with irrational leading coefficient. The system of functions $f(x+1), \ldots, f(x+s)$ is uniformly or not uniformly distributed in the s-dimensional unit cube depending on whether $s \leqslant n$ or $s > n$.*

Proof. Let us consider the function

$$F(x) = m_1 f(x+1) + \ldots + m_s f(x+s),$$

where $m_1, \ldots, m_s$ are integers not all zero. Using Taylor's formula, we obtain

$$f(x+\nu) = \sum_{j=0}^{n} \frac{1}{j!} f^{(j)}(x)\nu^j,$$

$$F(x) = \sum_{\nu=1}^{s} m_\nu f(x+\nu) = \sum_{j=0}^{n} \frac{1}{j!} f^{(j)}(x) \sum_{\nu=1}^{s} \nu^j m_\nu. \tag{283}$$

Since the determinant

$$\begin{vmatrix} 1 & 1 & \cdots & 1 \\ 1 & 2 & \cdots & s \\ \cdots & \cdots & \cdots & \cdots \\ \cdots & \cdots & \cdots & \cdots \\ 1 & 2^{s-1} & \cdots & s^{s-1} \end{vmatrix}$$

is not equal to zero, at least one of sums

$$\sum_{\nu=1}^{s} \nu^j m_\nu \qquad (j = 0, 1, \ldots, s-1) \tag{284}$$

does not vanish (otherwise the system of s linear homogeneous equations

$$\sum_{\nu=1}^{s} \nu^j m_\nu = 0 \qquad (j = 0, 1, \ldots, s-1)$$

would have the zero solution $m_1 = \ldots = m_s = 0$ only, which would contradict the choice of the quantities $m_1, \ldots, m_s$).

Denote by t the least value of j, under which the sum (284) does not equal zero:

$$\begin{aligned} &\sum_{\nu=1}^{s} \nu^j m_\nu = 0 \qquad (0 \leqslant j < t), \\ &\sum_{\nu=1}^{s} \nu^t m_\nu = \lambda \neq 0. \end{aligned} \tag{285}$$

Substituting these equalities into (283), we get

$$\begin{aligned} F(x) &= \sum_{j=t}^{n} \frac{1}{j!} f^{(j)} \sum_{\nu=1}^{s} \nu^j m_\nu \\ &= \frac{\lambda}{t!} f^{(t)}(x) + \sum_{j=t+1}^{n} \frac{1}{j!} f^{(j)}(x) \sum_{\nu=1}^{s} \nu^j m_\nu. \end{aligned} \tag{286}$$

Hence it is seen that the highest degree term of the polynomial $F(x)$ coincides with the highest degree term of the polynomial

$$\frac{\lambda}{t!} f^{(t)}(x) = C_n^t \lambda \alpha_n x^{n-1} + \ldots + \lambda \alpha_t.$$

Since λ is a nonzero integer and $t \leqslant s-1$, then under $s \leqslant n$ the function $F(x)$ is a polynomial of degree $n - t \geqslant 1$ with the irrational leading coefficient. But then fractional parts $\{F(x)\}$ are uniformly distributed by Theorem 26, and, therefore, the system of functions $f(x+1), \ldots, f(x+s)$ is uniformly distributed in the s-dimensional unit cube.

Let now $s = n + 1$. Consider consecutive finite differences with step being unity:

$$\Delta^{(1)} f(x+1) = f(x+2) - f(x+1),$$
$$\Delta^{(2)} f(x+1) = f(x+3) - 2f(x+2) + f(x+1),$$
$$\cdots\cdots\cdots\cdots\cdots\cdots\cdots\cdots\cdots\cdots$$
$$\Delta^{(n)} f(x+1) = f(x+n+1) - C_n^1 f(x+n) + \ldots \pm C_n^n f(x+1).$$

Since transition to a finite difference reduces the degree of a polynomial by unity, then $\Delta^{(1)} f(x+1)$ is a polynomial of degree $n-1$, $\Delta^{(2)} f(x+1)$ is a polynomial of degree $n-2$ and, finally, $\Delta^{(n)} f(x+1)$ is a constant. Therefore, with

$$m_\nu = (-1)^\nu C_n^{\nu-1} \qquad (\nu = 1, 2, \ldots, n+1)$$

we obtain

$$\left| \sum_{x=1}^{P} e^{2\pi i\, (m_1 f(x+1) + \ldots + m_{n+1} f(x+n+1))} \right| = \left| \sum_{x=1}^{P} e^{2\pi i\, \Delta^{(n)} f(x+1)} \right| = P.$$

But then by virtue of the multidimensional criterion of Weyl the system of functions $f(x+1), \ldots, f(x+s)$ is not uniformly distributed in the s-dimensional unit cube under $s = n+1$ (and, therefore, under any $s > n$ too). The theorem is proved completely.

By Theorem 27 there exist functions $f(x)$ such that the system of functions $f(x+1), \ldots, f(x+s)$ under s, which does not exceed a certain bound, is uniformly distributed in the s-dimensional unit cube. In the following theorem it is shown that there exist functions for which the restriction on the magnitude of s may be lifted.

A function $f(x)$ is called *completely uniformly distributed*, if for any $s \geqslant 1$ the system of functions

$$f(x+1), \ldots, f(x+s) \tag{287}$$

is uniformly distributed in the s-dimensional unit cube. It follows from (280) that a function $f(x)$ is completely uniformly distributed if and only if under every $s \geqslant 1$ and any choice of integers $m_1, \ldots, m_s$ not all zero the function

$$F(x) = m_1 f(x+1) + \ldots + m_s f(x+s) \tag{288}$$

is uniformly distributed.

THEOREM 28. *Under any $\alpha > 4$ a function $f(x)$ determined by the series*

$$f(x) = \sum_{k=0}^{\infty} e^{-k^\alpha} x^k$$

is completely uniformly distributed.

Proof. Let $m_1, \ldots, m_s$ be arbitrary integers not all zero and the function $F(x)$ be determined by the equality (288). Under $n \geqslant 2s$ we determine $Q(x)$ and $R(x)$ with the help of the equalities

$$Q(x) = \sum_{k=0}^{n} \alpha_k x^k, \qquad R(x) = \sum_{k=n+1}^{\infty} \alpha_k x^k,$$

where $\alpha_k = e^{-k^\alpha}$. Further, let

$$Q_s(x) = m_1 Q(x+1) + \ldots + m_s Q(x+s),$$
$$R_s(x) = m_1 R(x+1) + \ldots + m_s R(x+s).$$

Then, evidently, $f(x) = Q(x) + R(x)$ and

$$F(x) = m_1\big(Q(x+1) + R(x+1)\big) + \ldots + m_s\big(Q(x+s) + R(x+s)\big) = Q_s(x) + R_s(x).$$

By virtue of the multidimensional criterion of Weyl in order to prove the theorem it suffices to show that under any fixed positive integer s the estimate

$$\sum_{x=1}^{P} e^{2\pi i\, F(x)} = o(P)$$

is satisfied. Using Lemma 26, we get

$$\left|\sum_{x=1}^{P} e^{2\pi i\, F(x)}\right| = \left|\sum_{x=1}^{P} e^{2\pi i\, (Q_s(x) + R_s(x))}\right| \leqslant \left|\sum_{x=1}^{P} e^{2\pi i\, Q_s(x)}\right| + 2\pi \sum_{x=1}^{P} \big|R_s(x)\big|. \tag{289}$$

At first we shall estimate the magnitude of

$$R = \sum_{x=1}^{P} \big|R_s(x)\big|.$$

Determine n from the condition

$$n^{\alpha-1} \leqslant \log P < (n+1)^{\alpha-1}$$

and choose P in such a way that the inequality $n > \max(4ms,\ 2^{\alpha+1})$, where $m = \max_{1\leqslant\nu\leqslant s} |m_\nu|$, is satisfied. Then we obtain

$$R = \sum_{x=1}^{P}\left|\sum_{\nu=1}^{s} m_\nu R(x+\nu)\right| \leqslant \sum_{\nu=1}^{s}|m_\nu|\sum_{x=1}^{P} R(x+\nu)$$
$$\leqslant sm\sum_{x=1}^{R} R(x+s) = sm \sum_{k=n+1}^{\infty} e^{-k^\alpha}\sum_{x=1}^{P}(x+s)^k.$$

Hence, because of

$$\sum_{x=1}^{P}(x+s)^k \leqslant \sum_{x=1}^{P}\frac{(x+s+1)^{k+1}-(x+s)^{k+1}}{k+1} < \frac{(P+s+1)^{k+1}}{k+1},$$

it follows that

$$R \leqslant sm \sum_{k=n+1}^{\infty}\frac{e^{-k^\alpha}}{k+1}(P+s+1)^{k+1}. \tag{290}$$

Since by the determination of n

$$e^{\alpha(n+1)^{\alpha-1}} > e^{\alpha\log P} = P^\alpha,$$

then we get for the ratio of successive terms of the series (290)

$$\frac{(k+1)e^{-(k+1)^\alpha}(P+s+1)^{k+1}}{(k+2)e^{-k^\alpha}(P+s+1)^k} < \frac{P+s+1}{e^{\alpha k^{\alpha-1}}}$$
$$\leqslant \frac{P+s+1}{e^{\alpha(n+1)^{\alpha-1}}} < \frac{P+s+1}{P^\alpha} < \frac{1}{2},$$

and, therefore,

$$R \leqslant \frac{sm}{n+2}\,e^{-\alpha(n+1)^\alpha}(P+s+1)^{n+2}\sum_{j=1}^{\infty}\left(\frac{1}{2}\right)^{j-1}$$
$$= \frac{2\,sm}{n+2}\left(\frac{P+s+1}{e^{(n+1)^{\alpha-1}}}\right)^{n+1}(P+s+1)$$
$$< 4\,sm\left(\frac{P+s+1}{P}\right)^{n+1}\frac{P}{n+2} = o(P). \tag{291}$$

Now we shall estimate the sum

$$S = \sum_{x=1}^{P} e^{2\pi i\, Q_s(x)}.$$

Let, as above, $m = \max_{1\leqslant\nu\leqslant s}|m_\nu|$, $\alpha_j = e^{-j^\alpha}$ and t be determined by the equalities (285). Then

$$Q_s(x) = \sum_{\nu=1}^{s} m_\nu Q(x+\nu) = \sum_{\nu=1}^{s} m_\nu \sum_{k=0}^{n} \frac{1}{k!} Q^{(k)}(\nu) x^k = \sum_{k=0}^{n} \beta_k x^k,$$

where

$$\begin{aligned}\beta_k &= \sum_{\nu=1}^{s} \frac{1}{k!} Q^{(k)}(\nu) m_\nu \\ &= \sum_{j=k}^{n} C_j^k \left(\sum_{\nu=1}^{s} \nu^{j-k} m_\nu \right) \alpha_j = \sum_{j=k+t}^{n} C_j^k \left(\sum_{\nu=1}^{s} \nu^{j-k} m_\nu \right) e^{-j^\alpha}.\end{aligned}$$

Since, obviously,

$$\begin{aligned}\left| \sum_{j=k+t+1}^{n} C_j^k \left(\sum_{\nu=1}^{s} \nu^{j-k} m_\nu \right) e^{-j^\alpha} \right| &\leqslant ms\, e^{-(k+t+1)^\alpha} \sum_{j=t+1}^{n-k} C_{j+k}^j s^j \\ &< m(2s)^n e^{-(k+t+1)^\alpha},\end{aligned}$$

then under a certain θ_k ($|\theta_k| < 1$) we obtain

$$\begin{aligned}\beta_k &= C_{k+t}^k \left(\sum_{\nu=1}^{s} \nu^t m_\nu \right) e^{-(k+t)^\alpha} + \theta_k m(2s)^n e^{-(k+t+1)^\alpha} \\ &= \lambda C_{k+t}^t e^{-(k+t)^\alpha} + \theta_k m(2s)^n e^{-(k+t+1)^\alpha},\end{aligned} \tag{292}$$

where by (285) $1 \leqslant |\lambda| \leqslant ms^s$.

Determine r by the equality $r = [\frac{n}{2}] + 1$ and choose $k = r$. Since $0 \leqslant t < s$, $\alpha > 4$, and $n > \max(4ms,\ 2^{\alpha+1})$, then we have the estimates

$$(r+t+1)^\alpha > (r+t)^\alpha + \alpha(r+t)^{\alpha-1} > (r+t)^\alpha + \frac{1}{2}n^3,$$

$$m(2s)^n e^{-(r+t+1)^\alpha} < n^n e^{-n^3/2} e^{-(r+t)^\alpha} < \frac{1}{2} e^{-(r+t)^\alpha}$$

and by (292)

$$\frac{1}{2} e^{-(r+t)^\alpha} < \beta_r < (sn)^s e^{-(r+t)^\alpha},$$

$$(sn)^{-s} e^{\left(\frac{n+1}{2}\right)^\alpha} < \beta_r^{-1} < 2 e^{\left(\frac{n}{2}+s\right)^\alpha}. \tag{293}$$

Choose $q = [\beta_r^{-1}]$. Then, obviously, β_r may be written in the form

$$\beta_r = \frac{1}{q} + \frac{\theta}{q^2}, \qquad |\theta| \leqslant 1. \tag{294}$$

Let us show that

$$P \leqslant q \leqslant P^{r-1}. \tag{295}$$

In fact, since

$$1 + \left(\frac{n}{2} + s\right)^{\alpha} < \left(\frac{3}{4}n\right)^{\alpha} < \left[\frac{n}{2}\right] n^{\alpha-1} \leqslant (r-1)\log P,$$

then, using the inequality (293), we get

$$q \leqslant \beta_r^{-1} < e^{1+\left(\frac{n}{2}+s\right)^{\alpha}} \leqslant P^{r-1}.$$

On the other hand, from the evident inequalities

$$\log P \geqslant n^{\alpha-1} > s^2 n, \qquad (sn)^{-s} P > \left(\frac{e^{sn}}{sn}\right)^3 > 2,$$

$$\left(\frac{n+1}{2}\right)^{\alpha} > \frac{n}{2^{\alpha}}(n+1)^{\alpha-1} > 2 \log P$$

it follows by (293) that

$$q > \beta_r^{-1} - 1 > (sn)^{-s} e^{\left(\frac{n+1}{2}\right)^{\alpha}} - 1 > (sn)^{-s} P^2 - 1 > 2P - 1 \geqslant P.$$

The relations (294) and (295) show that for the sum

$$S = \sum_{x=1}^{P} e^{2\pi i\, Q_s(x)} = \sum_{x=1}^{P} e^{2\pi i\,(\beta_0 + \beta_1 x + \ldots + \beta_n x^n)}$$

the estimate obtained in Theorem 18 may be applied:

$$|S| \leqslant e^{3n} P^{1 - \frac{1}{24n^2 \log n}}.$$

Since $\alpha > 4$ and $n^{\alpha-1} \leqslant \log P < (n+1)^{\alpha-1}$, then

$$e^{3n} P^{-\frac{1}{24n^2 \log n}} < e^{3n - \frac{n^{\alpha-s}}{24 \log n}} \longrightarrow 0,$$

as $P \to \infty$, and, therefore, $S = o(P)$. But then by (289) and (291) we obtain the estimate

$$\left| \sum_{x=1}^{P} e^{2\pi i\, F(x)} \right| \leqslant |S| + 2\pi |R| = o(P)$$

equivalent to the theorem assertion.

Note. If a function $f(x)$ is completely uniformly distributed, then under any choice of positive integers t and r the system of functions

$$f(tx+1),\dots,f(tx+r) \tag{296}$$

is uniformly distributed in the r-dimensional unit cube.

Indeed, let $m_1,\dots,m_r$ be arbitrary integers not all zero. To prove uniform distribution of the system of functions (296) by the multidimensional criterion of Weyl it suffices to show that the sum

$$S=\sum_{x=1}^{P} e^{2\pi i\,F(tx)},$$

where $F(x)=m_1f(x+1)+\ldots+m_rf(x+r)$, has a nontrivial estimate $S=o(P)$. Using Lemma 2, we obtain

$$S=\sum_{x=1}^{tP} e^{2\pi i\,F(x)}\delta_t(x)=\frac{1}{t}\sum_{a=1}^{t}\sum_{x=1}^{tP} e^{2\pi i\left(F(x)+\frac{ax}{t}\right)},$$

$$|S|\leqslant\frac{1}{t}\sum_{a=1}^{t}\left|\sum_{x=1}^{tP} e^{2\pi i\left(F(x)+\frac{ax}{t}\right)}\right|. \tag{297}$$

Determine a function $F_a(x)$ by the equality $F_a(x)=F(x)+\frac{ax}{t}$ and denote by $\underset{h}{\Delta}F_a(x)$ its finite difference with step h:

$$\underset{h}{\Delta}F_a(x)=F_a(x+h)-F_a(x)=F(x+h)-F(x)+\frac{ah}{t}.$$

The difference $F(x+h)-F(x)$ is, obviously, a linear combination of consecutive values of the function $f(x)$:

$$\begin{aligned}F(x+h)-F(x)&=m_1\big(f(x+1+h)-f(x+1)\big)+\ldots\\&\quad+m_r\big(f(x+r+h)-f(x+r)\big)\\&=m_1'f(x+1)+\ldots+m_{r+h}'f(x+r+h),\end{aligned}$$

where $m_1',\dots,m_{r+h}'$ are integers not all zero. Hence, because the function $f(x)$ is completely uniformly distributed, by (288) the function $F(x+h)-F(x)$ is uniformly distributed. At the same time the function $\underset{h}{\Delta}F_a(x)$, which differs from $F(x+h)-F(x)$ by an additive constant only, is uniformly distributed as well. But then by Theorem 25 the function $F_a(x)$ is uniformly distributed too. Therefore, under any a from the interval $1\leqslant a\leqslant t$ we have

$$\sum_{x=1}^{tP} e^{2\pi i\left(F(x)+\frac{ax}{t}\right)}=\sum_{x=1}^{tP} e^{2\pi i\,F_a(x)}=o(P),$$

and it follows from (297) that

$$|S| \leqslant \frac{1}{t} \sum_{a=1}^{t} \left| \sum_{x=1}^{tP} e^{2\pi i\, F_a(x)} \right| = o(P).$$

The assertion (296) is proved.

§ 21. Normal and conjunctly normal numbers

Let $q \geqslant 2$ be an integer and α be an arbitrary number from the interval (0, 1). Let us write α by means of its q-adic expansion

$$\alpha = 0.\gamma_1\gamma_2 \cdots \gamma_x \cdots . \tag{298}$$

Denote by $N^{(P)}(\delta_1 \ldots \delta_n)$ the number of satisfactions of the equality

$$\gamma_{x+1} \cdots \gamma_{x+n} = \delta_1 \ldots \delta_n \qquad (x = 0, 1, \ldots, P-1), \tag{299}$$

where $\delta_1 \ldots \delta_n$ is an arbitrary fixed block of digits $\delta_\nu \in [0,\, q-1]$ and the equality (299) is considered as the equality of integers written by means of their q-adic expansion. As in § 8, $N^{(P)}(\delta_1 \ldots \delta_n)$ is equal, evidently, to the number of occurrences of the given block $\delta_1 \ldots \delta_n$ of digits of length n among the first P blocks

$$\gamma_1 \cdots \gamma_n,\ \gamma_2 \cdots \gamma_{n+1},\ \ldots,\ \gamma_P \cdots \gamma_{P+n-1}$$

formed by successive digits of the q-adic expansion (298) for α.

The number α is called *normal* to the base q, if for any fixed $n \geqslant 1$ under $P \to \infty$ the asymptotic equality

$$N^{(P)}(\delta_1 \ldots \delta_n) = \frac{1}{q^n} P + o(P)$$

holds.

The theory of normal numbers is closely connected with problems of uniform distribution of fractional parts of exponential functions αq^x. The following general lemma about uniform distribution of fractional parts of an arbitrary function $f(x)$ lies at the foundation of this connection.

LEMMA 31. *If there exists an infinite sequence of positive integers $m_1 < m_2 < \ldots < m_n < \ldots$ such that under every $n \geqslant 1$ and any integer ν with $0 \leqslant \nu \leqslant m_n - 1$ the number T_ν of satisfactions of the inequality*

$$\frac{\nu}{m_n} \leqslant \{f(x)\} < \frac{\nu+1}{m_n} \qquad (x = 1, 2, \ldots, P)$$

satisfies

$$T_\nu = \frac{1}{m_n} P + o(P) \tag{300}$$

as $P \to \infty$, then fractional parts of the function $f(x)$ are uniformly distributed.

Proof. Choose an arbitrary $\beta \in (0,\ 1]$ and denote by $N_P(\beta)$ the number of fractional parts $\{f(x)\}$ $(x = 1, 2, \ldots, P)$ falling into the interval $[0,\ \beta)$. Determine an integer b with the help of inequalities

$$\frac{b}{m_n} \leqslant \beta < \frac{b+1}{m_n}.$$

Then, obviously,

$$N_P\left(\frac{b}{m_n}\right) \leqslant N_P(\beta) \leqslant N_P\left(\frac{b+1}{m_n}\right).$$

Using the condition (300), we obtain

$$N_P\left(\frac{b}{m_n}\right) = \sum_{\nu=0}^{b-1} T_\nu = \frac{b}{m_n} P + o(P),$$

$$N_P\left(\frac{b+1}{m_n}\right) = \sum_{\nu=0}^{b} T_\nu = \frac{b+1}{m_n} P + o(P)$$

and, therefore,

$$\left(\frac{b}{m_n} - \beta\right) P + o(P) \leqslant N_P(\beta) - \beta P \leqslant \left(\frac{b+1}{m_n} - \beta\right) P + o(P),$$

$$\left|N_P(\beta) - \beta P\right| \leqslant \frac{1}{m_n} P + o(P).$$

Now let an arbitrarily small $\varepsilon > 0$ be given. Choose $n_0 = n_0(\varepsilon)$ so that for $n \geqslant n_0$ the inequality $\frac{1}{m_n} < \frac{\varepsilon}{2}$ is satisfied. Then, evidently,

$$\left|N_P(\beta) - \beta P\right| \leqslant \frac{\varepsilon}{2} P + o(P).$$

Hence under $P \geqslant P_0 = P_0(\varepsilon)$ we obtain

$$\left|N_P(\beta) - \beta P\right| \leqslant \varepsilon P$$

and, therefore,

$$\lim_{P \to \infty} \frac{1}{P} N_P(\beta) = \beta,$$

which is identical with the lemma assertion.

THEOREM 29. *A number α is normal to the base q if and only if fractional parts of the function αq^x are uniformly distributed.*

Proof. Choose an arbitrary block $\delta_1 \ldots \delta_n$ of digits with $0 \leqslant \delta_j \leqslant q-1$ and determine an integer ν with the help of the equality

$$0.\delta_1 \ldots \delta_n = \frac{\nu}{q^n}.$$

Let under a certain x the equality

$$\gamma_{x+1} \cdots \gamma_{x+n} = \delta_1 \ldots \delta_n \tag{301}$$

be fulfilled. Then

$$\{\alpha q^x\} = 0, \qquad \gamma_{x+1} \cdots \gamma_{x+n} + \frac{\theta_x}{q^n} = \frac{\nu + \theta_x}{q^n} \qquad (0 \leqslant \theta_x < 1),$$

and, therefore, the inequality

$$\frac{\nu}{q^n} \leqslant \{\alpha q^x\} < \frac{\nu+1}{q^n} \tag{302}$$

holds. It is also easy to verify that this inequality implies the equality (301). But then the inequality (302) and equation (301) under $x = 0, 1, \ldots, P-1$ have the same number of solutions being equal to $N^{(P)}(\delta_1 \ldots \delta_n)$. If fractional parts of the function αq^x are uniformly distributed, then under $x = 0, 1, \ldots, P-1$ for the number of satisfactions of the inequality (302) we get

$$N^{(P)}(\delta_1 \ldots \delta_n) = \frac{1}{q^n} P + o(P) \tag{303}$$

and, by definition, α is a normal number.

Now let α be a normal number to the base q. Then for any n and any block $\delta_1 \ldots \delta_n$ the equality (303) holds, and, therefore, under every integer ν $(0 \leqslant \nu < q)$ the number T_ν of satisfactions of the inequality

$$\frac{\nu}{q^n} \leqslant \{\alpha q^x\} < \frac{\nu+1}{q^n}$$

is asymptotically equal to $\frac{1}{q^n} P$:

$$T_\nu = \frac{1}{q^n} P + o(P).$$

Hence it follows by Lemma 31 that fractional parts of the function αq^x are uniformly distributed. The theorem is proved completely.

THEOREM 30. *Let $q \geqslant 2$ be an integer. Every number α determined by the equality*

$$\alpha = \sum_{k=1}^{\infty} \frac{[\theta_k q]}{q^k},$$

where $\theta_k = \{f(k)\}$ $(k = 1, 2, \ldots)$ are fractional parts of an arbitrary completely uniformly distributed function, is normal to the base q.

Proof. By the definition of completely uniform distribution under any fixed s the system of functions

$$f(x+1), \ldots, f(x+s) \tag{304}$$

is uniformly distributed in the s-dimensional unit cube. We split up every edge of the cube into q equal parts and, respectively, all the cube into q^s small cubes with the volume $\frac{1}{q^s}$. Then we enumerate the obtained small cubes, considering the number

$$\nu = \delta_{1\nu} q^{s-1} + \delta_{2\nu} q^{s-2} + \ldots + \delta_{s\nu},$$

where $\frac{\delta_{1\nu}}{q}, \ldots, \frac{\delta_{s\nu}}{q}$ are coordinates of the small cube vertex closest to the origin, as its serial number. Evidently, in this process the quantity ν takes on every integral value from 0 to $q^s - 1$.

It follows from uniform distribution of the system of functions (304), that under $x = 0, 1, \ldots, P-1$ the number N_ν of simultaneous fulfilment of inequalities

$$\frac{\delta_{j\nu}}{q} \leqslant \{f(x+j)\} < \frac{\delta_{j\nu} + 1}{q} \qquad (j = 1, 2, \ldots, s) \tag{305}$$

satisfies the relation

$$N_\nu = \frac{1}{q^s} P + o(P). \tag{306}$$

Since $\{f(x+j)\} = \theta_{x+j}$, the inequalities (305) are satisfied by those and only those x $(0 \leqslant x \leqslant P-1)$, for which

$$[\theta_{x+1} q] = \delta_{1\nu}, \ldots, [\theta_{x+s} q] = \delta_{s\nu}. \tag{307}$$

By virtue of the theorem condition

$$\begin{aligned} \{\alpha q^x\} &= \frac{[\theta_{x+1} q]}{q} + \ldots + \frac{[\theta_{x+s} q]}{q^s} + \frac{1}{q^s} \sum_{k=1}^{\infty} \frac{[\theta_{x+s+k} q]}{q^k} \\ &= \frac{[\theta_{x+1} q] q^{s-1} + \ldots + [\theta_{x+s} q]}{q^s} + \frac{\theta}{q^s} \qquad (0 < \theta < 1). \end{aligned}$$

Hence it is seen that under $\nu = \delta_{1\nu} q^{s-1} + \ldots + \delta_{s\nu}$ the number T_ν of fulfilments of the inequality

$$\frac{\nu}{q^s} \leqslant \{\alpha q^x\} < \frac{\nu + 1}{q^s} \qquad (x = 0, 1, \ldots, P-1) \tag{308}$$

coincides with the number of satisfactions of the equalities (307) and therefore coincides with N_ν. Since in the inequalities (308) ν may take on any value from the interval $0 \leqslant \nu \leqslant q^s - 1$, s is an arbitrary positive integer and by (306)

$$T_\nu = N_\nu = \frac{1}{q^s} P + o(P),$$

then Lemma 31 may be applied. By the lemma assertion fractional parts of the function αq^x are uniformly distributed, and therefore α is a normal number to the base q. The proof is completed.

Note. By Theorem 30 under any choice of a completely uniformly distributed function $f(x)$, a number α given by the series

$$\alpha = \sum_{k=1}^{\infty} \frac{[\{f(k)\}q]}{q^k} \tag{309}$$

is normal. It can be shown [39] that, conversely, every normal number to the base q is the sum of the series (309), where $\{f(k)\}$ are fractional parts of a certain completely uniformly distributed function. Thus a number α is normal to the base q if and only if digits of its q-adic expansion

$$\alpha = 0.\gamma_1\gamma_2 \cdots \gamma_x \cdots$$

satisfy the equality $\gamma_x = [\{f(x)\}q]$ $(x = 1, 2, \ldots)$, where $f(x)$ is a completely uniformly distributed function.

The notion of a normal number is naturally generalized for a case of several numbers. We consider numbers $\alpha_1, \ldots, \alpha_s$ given by their expansions to the bases $q_1, \ldots, q_s$, respectively:

$$\left.\begin{array}{l} \alpha_1 = 0.\gamma_1^{(1)} \ldots \gamma_k^{(1)} \ldots \\ \ldots\ldots\ldots\ldots\ldots\ldots \\ \alpha_s = 0.\gamma_1^{(s)} \ldots \gamma_k^{(s)} \ldots \end{array}\right\}. \tag{310}$$

Let $\delta_1^{(\nu)} \ldots \delta_n^{(\nu)}$ (or shortly Δ_ν) be arbitrary fixed blocks of n digits with respect to the base q_ν $(\nu = 1, 2, \ldots, s)$. Denote by $N^{(P)}(\Delta_1, \ldots, \Delta_s)$ the number of fulfilments of the system of equalities

$$\left.\begin{array}{l} \gamma_{x+1}^{(1)} \cdots \gamma_{x+n}^{(1)} = \delta_1^{(1)} \ldots \delta_n^{(1)} \\ \ldots\ldots\ldots\ldots\ldots\ldots\ldots\ldots \\ \gamma_{x+1}^{(s)} \cdots \gamma_{x+n}^{(s)} = \delta_1^{(s)} \ldots \delta_n^{(s)} \end{array}\right\}, \qquad x = 0, 1, \ldots, P-1, \tag{311}$$

considered as equalities of integers written in the scales of q_ν, respectively.

Numbers $\alpha_1, \ldots, \alpha_s$ are called *conjunctly normal* (to the bases $q_1, \ldots, q_s$), if under any choice of $\Delta_1, \ldots, \Delta_s$ the asymptotic equality

$$N^{(P)}(\Delta_1, \ldots, \Delta_s) = \frac{1}{q_1^n \ldots q_s^n} P + o(P)$$

holds. Thus, the numbers $\alpha_1, \ldots, \alpha_s$ are conjunctly normal, if under any choice of digits $\delta_j^{(\nu)}$ $(j = 1, 2, \ldots, n, \ \nu = 1, 2, \ldots, s)$ every of $q_1^n \ldots q_s^n$ possible distinct blocks of digits

$$\left.\begin{array}{c} \delta_1^{(1)} \ldots \delta_n^{(1)} \\ \ldots\ldots\ldots \\ \delta_1^{(s)} \ldots \delta_n^{(s)} \end{array}\right\} \qquad (0 \leqslant \delta_j^{(\nu)} \leqslant q_\nu - 1)$$

occurs among the blocks

$$\left.\begin{array}{c} \gamma_{x+1}^{(1)} \ldots \gamma_{x+n}^{(1)} \\ \ldots\ldots\ldots\ldots \\ \gamma_{x+1}^{(s)} \ldots \gamma_{x+n}^{(s)} \end{array}\right\} \qquad (x = 0, 1, \ldots),$$

formed by successive digits of the expansions (310) with asymptotically equal frequency. In the same way as in Theorem 29, it is easy to show that the numbers $\alpha_1, \ldots, \alpha_s$ are conjunctly normal if and only if the system of functions

$$\alpha_1 q_1^x, \ldots, \alpha_s q_s^x \tag{312}$$

is uniformly distributed in the s-dimensional unit cube.

A connection between conjunctly normal numbers and completely uniformly distributed functions is established in the following theorem generalizing Theorem 30.

THEOREM 31. *Let $q_1, \ldots, q_s$ be integers greater than unity and $f(x)$ be an arbitrary completely uniformly distributed function. Then numbers $\alpha_1, \ldots, \alpha_s$, determined by the equalities*

$$\alpha_\nu = \sum_{k=1}^{\infty} \frac{[\{f(sk + \nu)\} q_\nu]}{q_\nu^k} \qquad (\nu = 1, 2, \ldots, s), \tag{313}$$

are conjunctly normal.

Proof. We determine quantities $\gamma_k^{(\nu)}$ by the equalities

$$f(sx + s + 1), \ \ldots, \ f(sx + s + ns).$$

Obviously, $\gamma_k^{(\nu)}$ are integers from the interval $[0,\, q_\nu - 1]$. Therefore, the series (313) may be considered as q_ν-adic expansion of the numbers α_ν, and we may write instead of (313):

$$\left.\begin{array}{l} \alpha_1 = 0.\gamma_1^{(1)} \dots \gamma_k^{(1)} \dots \\ \dots\dots\dots\dots\dots\dots \\ \alpha_s = 0.\,\gamma_1^{(s)} \dots \gamma_k^{(s)} \dots \end{array}\right\}.$$

Let $\delta_1^{(\nu)} \dots \delta_n^{(\nu)}$ be arbitrary fixed blocks of digits with respect to the base q_ν, and, as in (311), $N^{(P)}(\Delta_1, \dots, \Delta_s)$ be the number of fulfilments of the system of equalities

$$\left.\begin{array}{l} \gamma_{x+1}^{(1)} \cdots \gamma_{x+n}^{(1)} = \delta_1^{(1)} \dots \delta_n^{(1)} \\ \dots\dots\dots\dots\dots\dots\dots\dots \\ \gamma_{x+1}^{(s)} \cdots \gamma_{x+n}^{(s)} = \delta_1^{(s)} \dots \delta_n^{(s)} \end{array}\right\}, \qquad x = 0, 1, \dots, P-1. \tag{314}$$

The equalities (314) are, evidently, equivalent to the equalities

$$\gamma_{x+1}^{(\nu)} = \delta_1^{(\nu)}, \;\dots,\; \gamma_{x+n}^{(\nu)} = \delta_n^{(\nu)},$$

$$\nu = 1, 2, \dots, s, \qquad x = 0, 1, \dots, P-1.$$

Using the determination of the quantities $\gamma_k^{(\nu)}$, we rewrite these equalities in the form

$$\left.\begin{array}{l} [\{f(sx + s + \nu)\} q_\nu] = \delta_1^{(\nu)} \\ \dots\dots\dots\dots\dots\dots\dots\dots \\ [\{f(sx + ns + \nu)\} q_\nu] = \delta_n^{(\nu)} \end{array}\right\}, \qquad \begin{array}{l} \nu = 1, 2, \dots, s, \\ x = 0, 1, \dots, P-1. \end{array}$$

In turn these equalities are equivalent to the system of inequalities

$$\left.\begin{array}{l} \dfrac{\delta_1^{(\nu)}}{q_\nu} \leqslant \{f(sx + s + \nu)\} < \dfrac{\delta_1^{(\nu)} + 1}{q_\nu} \\ \dots\dots\dots\dots\dots\dots\dots\dots\dots\dots \\ \dfrac{\delta_n^{(\nu)}}{q_\nu} \leqslant \{f(sx + ns + \nu)\} < \dfrac{\delta_n^{(\nu)} + 1}{q_\nu} \end{array}\right\}, \qquad \begin{array}{l} \nu = 1, 2, \dots, s, \\ x = 0, 1, \dots, P-1. \end{array} \tag{315}$$

Thus $N^{(P)}(\Delta_1, \dots, \Delta_s)$ is equal to the number of solutions of the system (315), i.e., to the number of fractional parts of the system of functions

$$f(sx + s + 1), \;\dots,\; f(sx + s + ns) \tag{316}$$

falling into the ns-dimensional cuboid determined by the inequalities (315). This cuboid lies inside the sn-dimensional unit cube; its volume does not depend upon the choice of the quantities $\delta_1^{(\nu)}, \ldots, \delta_n^{(\nu)}$ and is equal to $\frac{1}{q_1^n \cdots q_s^n}$.

By the note (296) the system of functions $f(sx+1), \ldots, f(sx+ns)$ is uniformly distributed in the ns-dimensional unit cube. But then the system of functions (316) is uniformly distributed too. Therefore the asymptotic equality

$$N^{(P)}(\Delta_1, \ldots, \Delta_s) = \frac{1}{q_1^n \cdots q_s^n} P + o(P)$$

holds. Hence, using the definition of conjunct normality, we get the theorem assertion.

§ 22. Distribution of digits in period part of periodical fractions

A problem of distribution of digits in the complete period of fractions, arising in expansion of rational numbers with respect to an arbitrary base, was considered in § 8. We shall keep notation introduced there and suppose that $q \geqslant 2$, $m \equiv 1 \pmod 2$, $(q, m) = 1$, $(a, m) = 1$, and τ is a period of a q-adic expansion of the number $\frac{a}{m}$:

$$\frac{a}{m} = \left[\frac{a}{m}\right] + 0.\gamma_1\gamma_2\cdots\gamma_x\cdots, \qquad \gamma_{x+\tau} = \gamma_x \quad (x \geqslant 1). \tag{317}$$

We let $N_m^{(P)}(\delta_1 \ldots \delta_n)$ denote the number of occurrences of a given block $\delta_1 \ldots \delta_n$ of digits of length n among the first P blocks

$$\gamma_1 \cdots \gamma_n, \gamma_2 \cdots \gamma_{n+1}, \ldots, \gamma_P \cdots \gamma_{P+n-1}$$

formed by successive digits of the expansion (317). Under $P \leqslant \tau$ we determine $R_n(P)$ by the equality

$$N_m^{(P)}(\delta_1 \ldots \delta_n) = \frac{1}{q^n} P + R_n(P).$$

A degree of uniformity of distribution of digit blocks $\delta_1 \ldots \delta_n$ in the period or in a part of the period can be, evidently, characterized by an estimate for deviation of $N_m^{(P)}(\delta_1 \ldots \delta_n)$ from the average value $\frac{1}{q^n}P$, i.e., by an estimate for the quantity $R_n(P)$.

The question concerning distribution of digit blocks in a part of the period (under $P < \tau$) is being solved in different ways depending on whether P is greater or less than $m^{\frac{1}{2}+\varepsilon}$, where ε is an arbitrarily small positive number.

Let $m = p_1^{\alpha_1} \dots p_s^{\alpha_s}$ be the prime factorization of odd m, τ and τ_1 be the orders of q for moduli m and $p_1 \dots p_s$, respectively. Determine quantities $\beta_1, \dots, \beta_s$ by the help of the conditions

$$q^{\tau_1} - 1 = u_0 p_1^{\beta_1} \dots p_s^{\beta_s} \qquad (u_0, p_1 \dots p_s) = 1. \tag{318}$$

Without loss of generality, it may be assumed that $\beta_\nu < \alpha_\nu$ under $\nu = 1, 2, \dots, r$ and $\alpha_\nu \leqslant \beta_\nu$ under $\nu = r+1, \dots, s$. Obviously, under $m_1 = p_1^{\beta_1} \dots p_r^{\beta_r} \, p_{r+1}^{\alpha_{r+1}} \dots p_s^{\alpha_s}$ the order of q for modulus m_1 is equal to τ_1 and by (96) $m_1 \tau = m \tau_1$.

The distribution of digits in a large part of the period (for $P > m^{\frac{1}{2}+\varepsilon}$) is investigated rather easily with the help of the following lemma.

LEMMA 32. *Let $q \geqslant 2$, the quantities m, m_1, τ, τ_1 be determined according to (318), b be an arbitrary positive integer, and $d = (b, m)$. Then under $d < \frac{m}{m_1}$ for every $P \leqslant \tau$ we have the estimate*

$$\left| \sum_{x=0}^{P-1} e^{2\pi i \frac{bq^x}{m}} \right| \leqslant \sqrt{\frac{m}{d}} \left(1 + \log \frac{m}{d} \right). \tag{319}$$

Proof. At first we shall consider a case $d = 1$. As it was shown in § 7, under any integer c the estimate

$$\left| \sum_{x=0}^{\tau-1} e^{2\pi i \left(\frac{bq^x}{m} + \frac{cx}{\tau} \right)} \right| \leqslant \sqrt{m}$$

holds. Therefore using Theorem 2, we obtain the estimate

$$\begin{aligned} \left| \sum_{x=0}^{P-1} e^{2\pi i \frac{bq^x}{m}} \right| &\leqslant \max_{1 \leqslant c \leqslant \tau} \left| \sum_{x=0}^{\tau-1} e^{2\pi i \left(\frac{bq^x}{m} + \frac{cx}{\tau} \right)} \right| (1 + \log \tau) \\ &\leqslant \sqrt{m} \, (1 + \log m), \end{aligned} \tag{320}$$

coinciding with the estimate (319) under $d = 1$.

Let now $d > 1$. Since $m = p_1^{\alpha_1} \dots p_s^{\alpha_s}$ and $d \backslash m$, then d may be written in the form $d = p_1^{k_1} \dots p_s^{k_s}$, where $0 \leqslant k_1 \leqslant \alpha_1, \dots, 0 \leqslant k_s \leqslant \alpha_s$. Note that the inequality $k_\nu < \alpha_\nu - \beta_\nu$ is satisfied by at least one quantity k_ν under $\nu = 1, 2, \dots, r$. Indeed, otherwise

$$d \geqslant p_1^{\alpha_1 - \beta_1} \dots p_r^{\alpha_r - \beta_r} = \frac{m}{m_1},$$

which contradicts the theorem hypothesis. But then by Theorem 9 we have the equality

$$\sum_{x=0}^{\tau-1} e^{2\pi i \frac{bq^x}{m}} = 0. \tag{321}$$

Write down b and m in the form $b = b'd,\ m = m'd,\ (b', m') = 1$ and denote by τ' the order to which q belongs for modulus m'. Further let P' and Q be determined by the conditions

$$P = Q\tau' + P', \qquad 0 \leqslant P' < \tau'.$$

Then, using the property of complete sums, we obtain

$$\sum_{x=0}^{\tau-1} e^{2\pi i \frac{bq^x}{m}} = \frac{\tau}{\tau'} \sum_{x=0}^{\tau'-1} e^{2\pi i \frac{b'q^x}{m'}}$$

and by (321)

$$\sum_{x=0}^{\tau'-1} e^{2\pi i \frac{b'q^x}{m'}} = 0.$$

Therefore,

$$\sum_{x=0}^{P-1} e^{2\pi i \frac{bq^x}{m}} = \sum_{y=0}^{Q-1} \sum_{z=0}^{\tau'-1} e^{2\pi i \frac{b'q^{\tau'_y+z}}{m'}} + \sum_{z=0}^{P'-1} e^{2\pi i \frac{b'q^{\tau_1 Q+z}}{m'}}$$
$$= Q \sum_{z=0}^{\tau'-1} e^{2\pi i \frac{b'q^z}{m'}} + \sum_{z=0}^{P'-1} e^{2\pi i \frac{b'q^z}{m'}} = \sum_{z=0}^{P'-1} e^{2\pi i \frac{b'q^z}{m'}}.$$

Hence, using the estimate (320), we get the lemma assertion:

$$\left| \sum_{x=0}^{P-1} e^{2\pi i \frac{bq^x}{m}} \right| = \left| \sum_{x=0}^{P'-1} e^{2\pi i \frac{b'q^x}{m'}} \right|$$
$$\leqslant \sqrt{m'}\,(1 + \log m') = \sqrt{\frac{m}{d}} \left(1 + \log \frac{m}{d}\right).$$

THEOREM 32. *Let $P \leqslant \tau$, $m = p_1^{\alpha_1} \dots p_s^{\alpha_s}$, $\alpha_\nu \geqslant 2\ (\nu = 1, 2, \dots, s)$, and the quantity $N_m^{(P)}(\delta_1 \dots \delta_n)$ be determined by (317). Then for every $n \geqslant 1$ under any choice of digits $\delta_1 \dots \delta_n$ and any $\varepsilon > 0$ we have the equality*

$$N_m^{(P)}(\delta_1 \dots \delta_n) = \frac{1}{q^n} P + O\left(m^{\frac{1}{2}+\varepsilon}\right),$$

where the constant implied by the symbol "O" depends on ε only.

Proof. We determine integers t, b, and h as in § 8:

$$0.\delta_1 \dots \delta_n = \frac{t}{q^n}, \qquad b = \left[\frac{tm}{q^n}\right], \qquad h = \left[\frac{(t+1)m}{q^n}\right] - \left[\frac{tm}{q^n}\right], \tag{322}$$

and denote by $T_m^{(P)}(b,h)$ the number of solutions of the congruence

$$aq^x \equiv y + b \pmod m, \qquad 0 \leqslant x < P, \quad 1 \leqslant y \leqslant h.$$

Then by Lemma 10

$$N_m^{(P)}(\delta_1 \dots \delta_n) = T_m^{(P)}(b,h). \tag{323}$$

Using Lemma 2, we obtain

$$T_m^{(P)}(b,h) = \sum_{x=0}^{P-1} \sum_{y=1}^{h} \delta_m\left(aq^x - y - b\right) = \frac{h}{m} P + R, \tag{324}$$

where R is determined by the equality

$$\begin{aligned} R &= \sum_{x=0}^{P-1} \sum_{y=1}^{h} \left(\delta_m\left(aq^x - y - b\right) - \frac{1}{m}\right) \\ &= \frac{1}{m} \sum_{z=1}^{m-1} \left(\sum_{y=1}^{h} e^{-2\pi i \frac{(y+b)z}{m}} \right) \left(\sum_{x=0}^{P-1} e^{2\pi i \frac{azq^x}{m}} \right). \end{aligned}$$

Let us estimate the quantity $|R|$. Obviously,

$$\begin{aligned} |R| &\leqslant \frac{1}{m} \sum_{z=1}^{m-1} \left| \sum_{y=1}^{h} e^{2\pi i \frac{zy}{m}} \right| \left| \sum_{x=0}^{P-1} e^{2\pi i \frac{azq^x}{m}} \right| \\ &\leqslant \frac{1}{m} \sum_{z=1}^{m-1} \frac{1}{2\left\| \frac{z}{m} \right\|} \left| \sum_{x=0}^{P-1} e^{2\pi i \frac{azq^x}{m}} \right| \leqslant \sum_{z=1}^{m-1} \frac{1}{z} \left| \sum_{x=0}^{P-1} e^{2\pi i \frac{azq^x}{m}} \right|. \end{aligned} \tag{325}$$

Let m_1 be determined by (318) and d be the greatest common divisor of m and z. Using under $d \geqslant \frac{m}{m_1}$ the trivial estimation and applying under $d < \frac{m}{m_1}$ the estimate

$$\left| \sum_{x=0}^{P-1} e^{2\pi i \frac{azq^x}{m}} \right| \leqslant \sqrt{m}\,(1 + \log m),$$

following from Lemma 32, from (325) we get

$$\begin{aligned} |R| &\leqslant P \sum_{(z,m) \geqslant m m_1^{-1}} \frac{1}{z} + \sqrt{m}\,(1 + \log m) \sum_{z=1}^{m-1} \frac{1}{z} \\ &\leqslant P \frac{m}{m_1} \sum_{d \backslash m} 1 + \sqrt{m}\,(1 + \log m)^2. \end{aligned} \tag{326}$$

Since by the hypothesis $\alpha_\nu \geqslant 2$ $(\nu = 1, 2, \ldots, s)$, then

$$p_1 \ldots p_s \leqslant p_1^{\frac{\alpha_1}{2}} \ldots p_s^{\frac{\alpha_s}{s}} = \sqrt{m}\,.$$

But then for the order of q for modulus $p_1 \ldots p_s$ the estimate $\tau_1 < p_1 \ldots p_s \leqslant \sqrt{m}$ holds and, therefore,

$$P\frac{m_1}{m} \leqslant \tau\frac{m_1}{m} = \tau_1 \leqslant \sqrt{m}\,.$$

Observing that under any $\varepsilon > 0$

$$\sum_{d\backslash m} 1 \leqslant C(\varepsilon)m^\varepsilon,$$

we obtain from (326)

$$|R| \leqslant C(\varepsilon)m^{\frac{1}{2}+\varepsilon} + \sqrt{m}\,(1 + \log\, m)^2 = O\Big(m^{\frac{1}{2}+\varepsilon}\Big).$$

Now it follows from (324) that

$$T_m^{(P)}(b,\, h) = \frac{h}{m}P + O\Big(m^{\frac{1}{2}+\varepsilon}\Big).$$

Hence by (322) and (323) we get the assertion of the theorem:

$$N_m^{(P)}(\delta_1 \ldots \delta_n) = \frac{1}{m}\left[\frac{m}{q^n} + \left\{\frac{tm}{q^n}\right\}\right]P + R = \frac{1}{q^n}P + O\Big(m^{\frac{1}{2}+\varepsilon}\Big).$$

Note that the uniformity of distribution of digit blocks $\delta_1 \ldots \delta_n$ in a part of the period of the fraction $\frac{a}{m}$ follows from Theorem 32 only if P belongs to the interval $m^{\frac{1}{2}+\varepsilon} < P < \tau$, i.e., if the period is sufficiently large and a sufficiently large part of it is considered. It is so, for example, if we require the fulfilment of the inequalities $\alpha_\nu > 2\beta_\nu$ $(\nu = 1, 2, \ldots, s)$.

Indeed, in this case by (318) $m_1 = p_1^{\beta_1} \ldots p_s^{\beta_s}$. Since $m = p_1^{\alpha_1} \ldots p_s^{\alpha_s}$ and $\tau = \frac{m\tau_1}{m_1}$, then

$$\tau \geqslant \frac{m}{m_1} = p_1^{\alpha_1-\beta_1} \ldots p_s^{\alpha_s-\beta_s} \geqslant p_1^{\frac{\alpha_1+1}{2}} \ldots p_s^{\frac{\alpha_s+1}{2}}. \tag{327}$$

Let $\alpha = \max_{1\leqslant\nu\leqslant s} \alpha_\nu$. Then

$$p_1 \ldots p_s \geqslant \left(p_1^{\alpha_1} \ldots p_s^{\alpha_s}\right)^{\frac{1}{\alpha}} = m^{\frac{1}{\alpha}}$$

and, using the estimate (327), we obtain the required bound for the magnitude of the period:

$$\tau \geqslant \left(mp_1 \ldots p_s\right)^{\frac{1}{2}} \geqslant m^{\frac{1}{2}+\frac{1}{2\alpha}}.$$

Now we start on the question concerning the distribution of digit blocks in a small part of the period. This question is more difficult than the former and more complicated methods of the estimation of corresponding exponential sums have to be invoked for its solution. We restrict ourselves to the case $m = p^\alpha$, where $p > 2$ is a prime. We assume that the quantities τ, τ_1, and β are chosen, as before, by (318).

THEOREM 33. *Let* $(q,p) = 1$, $(a,p) = 1$, $\alpha > 16\beta$, *and* r *be determined by the equality* $P^r = p^\alpha$. *If* $2 \leqslant r < \frac{\alpha}{8\beta}$, *then we have the estimate*

$$\left|\sum_{x=0}^{P-1} e^{2\pi i \frac{aq^x}{p^\alpha}}\right| < 3P^{1-\frac{\gamma}{r^2}},$$

where $\gamma = \frac{1}{2\cdot 10^6}$.

Proof. If $P \leqslant e^{36r}$, then the theorem assertion is trivial, because of

$$3P^{1-\frac{\gamma}{r^2}} \geqslant 3Pe^{-\frac{36\gamma}{r}} \geqslant 3Pe^{-18\gamma} > P.$$

Let $P > e^{36r}$. We determine integers s and n with the help of the conditions

$$s \leqslant \frac{\alpha}{4r} < s+1, \qquad n < \frac{\alpha}{s} \leqslant n+1. \tag{328}$$

It is easy to verify, that the estimates

$$s > \beta, \qquad p^s < P^{\frac{1}{4}}, \qquad 7 \leqslant n < s(p-1)$$

holds.

In fact, since by the hypothesis $\alpha > 8\beta r$ and $p^\alpha = P^r$, then we obtain from (328)

$$s > \frac{\alpha}{4r} - 1 > 2\beta - 1 \geqslant \beta, \qquad p^s \leqslant p^{\frac{\alpha}{4r}} = P^{\frac{1}{4}}.$$

Further, evidently, $n \geqslant \frac{\alpha}{s} - 1 \geqslant 4r - 1 \geqslant 7$ and, finally,

$$n < \frac{\alpha}{s} < \frac{4r(s+1)}{s} \leqslant 6r < \frac{1}{6}\log P = \frac{\alpha \log p}{6r} < s\log p < s(p-1).$$

Determine integers $a_1, \ldots, a_n$ by means of the equality

$$n!\left(C_x^1 up^s + \ldots + C_x^n u^n p^{sn}\right) = a_1 p^s x + \ldots + a_n p^{sn} x^n \tag{329}$$

and show that under $(u,p) = 1$ there are no multiples of p^s among the quantities a_ν $(\nu = 1, 2, \ldots, n)$. Indeed, by comparing the coefficients of x^ν, we get

$$a_\nu p^{s\nu} \equiv \frac{n!}{\nu!} u^\nu p^{s\nu} \pmod{p^{s(\nu+1)}}, \qquad a_\nu \equiv \frac{n!}{\nu!} u^\nu \pmod{p^s}.$$

Let p^{ω_ν} be the highest power of p dividing $\frac{n!}{\nu!}$. Then because of $n < s(p-1)$ we obtain

$$\omega_\nu \leqslant \left[\frac{n}{p}\right] + \left[\frac{n}{p^2}\right] + \ldots < \frac{n}{p-1} < s,$$

and, therefore,

$$(a_\nu,\, p^s) = p^{\omega_\nu}, \qquad 0 \leqslant \omega_\nu < s, \quad \omega_n = 0. \tag{330}$$

Denote by τ_s the order of q for modulus p^s. Since $s > \beta$, then by (96)

$$q^{\tau_s} = 1 + up^s, \qquad (u,\, p) = 1, \quad \tau_s = \tau_1 p^{s-\beta} < p^s.$$

Using $s\,(n+1) \geqslant \alpha$, under any integer $x \geqslant 0$ we get

$$\begin{aligned} q^{\tau_s x} &= \left(1 + up^s\right)^x = 1 + C_x^1 up^s + C_x^2 u^2 p^{2s} + \ldots \\ &\equiv 1 + C_x^1 up^s + \ldots + C_x^n u^n p^{sn} \pmod{p^\alpha}. \end{aligned}$$

Hence according to Lemma 19 by virtue of (328) and (329) it follows that

$$\begin{aligned} \left| \sum_{x=0}^{P-1} e^{2\pi i \frac{aq^x}{p^\alpha}} \right| &\leqslant \frac{1}{p^{2s}} \sum_{x=0}^{P-1} \left| \sum_{y,z=1}^{p^s} e^{2\pi i \frac{aq^{x+\tau_s yz}}{p^\alpha}} \right| + 2\tau_s p^{2s} \\ &\leqslant \frac{1}{p^{2s}} \sum_{x=0}^{P-1} \left| \sum_{y,z=1}^{p^s} e^{2\pi i \frac{aq^x (C_{yz}^1 up^s + \ldots + C_{yz}^n u^n p^{sn})}{p^\alpha}} \right| + 2p^{3s} \\ &\leqslant \frac{1}{p^{2s}} \sum_{x=0}^{P-1} \left| \sum_{y,z=1}^{p^s} e^{2\pi i \frac{aq^x (a_1 p^s yz + \ldots + a_n p^{sn} y^n z^n)}{n! p^\alpha}} \right| + 2P^{\frac{3}{4}}. \end{aligned} \tag{331}$$

We determine a function $f_x(y,\, z)$ and integers $b_\nu,\ q_\nu$ with the help of the conditions

$$\begin{aligned} f_x(y,\, z) &= \frac{aq^x(a_1 p^s yz + \ldots + a_n p^{sn} y^n z^n)}{n!\, p^\alpha}, \\ \frac{aa_\nu q^x p^{s\nu}}{n!\, p^\alpha} &= \frac{b_\nu}{q_\nu}, \qquad (b_\nu,\, q_\nu) = 1 \quad (\nu = 1, 2, \ldots, n). \end{aligned} \tag{332}$$

Then we obtain from (331)

$$\left| \sum_{x=0}^{P-1} e^{2\pi i \frac{aq^x}{p^\alpha}} \right| \leqslant \frac{1}{p^{2s}} \sum_{x=0}^{P-1} \left| \sum_{y,z=1}^{p^s} e^{2\pi i\, f_x(y,z)} \right| + 2\,P^{\frac{3}{4}}, \tag{333}$$

where by (330) and (332)

$$f_x(y,\, z) = \frac{b_1}{q_1}\, yz + \ldots + \frac{b_n}{q_n}\, y^n z^n,$$

$$(b_\nu,\, q_\nu) = 1, \qquad p^{\alpha - \nu s - \omega_\nu} \leqslant q_\nu \leqslant n!\, p^{\alpha - \nu s - \omega_\nu} \quad (\nu = 1, 2, \ldots, n). \tag{334}$$

To estimate the sum

$$\sigma_x = \sum_{y,z=1}^{p^s} e^{2\pi i\, f_x(y,z)}$$

we shall use the corollary of Lemma 25

$$\left|\sigma_x\right|^{2k^2} = \left|\sum_{y,z=1}^{p^s} e^{2\pi i\left(\frac{b_1}{q_1}yz+\ldots+\frac{b_n}{q_n}y^n z^n\right)}\right|^{2k^2}$$
$$\leqslant (2k)^{2n}\left(p^s\right)^{4k^2-2k+\frac{1}{2k}} N_k(p^s)\prod_{\nu=1}^{n}\min\left(p^{s\nu},\, \sqrt{q_\nu}+\frac{p^{s\nu}}{\sqrt{q_\nu}}\right). \qquad (335)$$

Since $sn < \alpha \leqslant s(n+1)$ and $0 \leqslant \omega_\nu < s$, it follows from (334) that under $\frac{n+1}{2} \leqslant \nu < n$ the estimates

$$q_\nu \leqslant n!\,p^{s(n+1-\nu)} \leqslant n!\,p^{s\nu}, \qquad q_\nu > p^{s(n-1-\nu)},$$
$$\frac{1}{\sqrt{q_\nu}}+\frac{\sqrt{q_\nu}}{p^{s\nu}} \leqslant \frac{1+n!}{\sqrt{q_\nu}} < n^n p^{-\frac{s}{2}(n-1-\nu)}$$

hold. But then

$$\prod_{\nu=1}^{n}\min\left(p^{s\nu},\, \sqrt{q_\nu}+\frac{p^{s\nu}}{\sqrt{q_\nu}}\right) = p^{s\frac{n(n+1)}{2}}\prod_{\nu=1}^{n}\min\left(1,\, \frac{\sqrt{q_\nu}}{p^{s\nu}}+\frac{1}{\sqrt{q_\nu}}\right)$$
$$\leqslant p^{s\frac{n(n+1)}{2}}\prod_{\frac{n}{2}<\nu<n} n^n p^{-s\frac{n-1-\nu}{2}} < n^{n^2} p^{s\frac{n(n+1)}{2}-s\frac{(n-2)(n-4)}{16}}. \qquad (336)$$

To estimate $N_k(p^s)$ we choose $k = \frac{n(n+1)}{2} + 8n^2$ and use Theorem 16

$$N_k(p^s) \leqslant (2\,k)^{2k}(2\,n)^{n^3} p^{2sk-s\frac{n(n+1)}{2}+s\frac{n(n-1)}{2}\left(1-\frac{1}{n}\right)^{8n}}$$
$$\leqslant (2n)^{4n^3} p^{2sk-s\frac{n(n+1)}{2}+s\frac{n(n-1)}{5950}}. \qquad (337)$$

Now it follows from (335) by (336) and (337) that

$$\left|\sigma_x\right|^{2k^2} \leqslant (2k)^{2n} n^{n^2}(2n)^{4n^3}\left(p^s\right)^{4k^2+\frac{1}{2k}+\frac{n(n-1)}{5950}-\frac{(n-2)(n-4)}{16}}$$
$$\leqslant (2n)^{5n^3} p^{4sk^2-\frac{sn^2}{56}},$$
$$\left|\sigma_x\right| \leqslant 2p^{2s-\frac{s}{9100n^2}}.$$

Since $\alpha n^2 < (\frac{\alpha}{s})^3 s < (6r)^3 s$, then

$$p^{\frac{s}{n^2}} = P^{\frac{rs}{\alpha n^2}} > P^{\frac{1}{216 r^2}},$$

and, therefore,

$$\left|\sigma_x\right| < 2p^{2s} P^{-\frac{1}{2\cdot 10^6 r^2}} = 2p^{2s} P^{-\frac{\gamma}{r^2}},$$

where $\gamma = \frac{1}{2\cdot 10^6}$. Substituting this estimate into (333), we obtain the theorem assertion:

$$\left|\sum_{x=0}^{P-1} e^{2\pi i \frac{aq^x}{p^\alpha}}\right| \leqslant \frac{1}{p^{2s}} \sum_{x=0}^{P-1} \left|\sigma_x\right| + 2P^{\frac{3}{4}} \leqslant 2P^{1-\frac{\gamma}{r^2}} + 2P^{\frac{3}{4}} \leqslant 3P^{1-\frac{\gamma}{r^2}}.$$

Note that the obtained estimate is nontrivial starting with values P, which are very small with respect to the period τ. In particular, under any $\varepsilon > 0$ and sufficiently large α for $P > \tau^\varepsilon$ we have the estimate

$$\left|\sum_{x=0}^{P-1} e^{2\pi i \frac{aq^x}{p^\alpha}}\right| < 3P^{1-\gamma_1 \varepsilon^2},$$

where γ_1 is a certain positive constant. It is also easy to verify that the least values P, from which the estimate of Theorem 33 is nontrivial, have the order $e^{c \log^{\frac{2}{3}} r}$, where c is a certain absolute constant.

Let, as before, $N_m^{(P)}(\delta_1 \ldots \delta_n)$ be the number of occurrences of a given digit block $\delta_1 \ldots \delta_n$ among the first P blocks formed by successive digits of the q-adic expansion of the fraction $\frac{a}{m}$.

THEOREM 34. *If* $p > 2$ *is a prime,* $(a, p) = 1$, $(q, p) = 1$, $P^r = p^\alpha$, *and* $3 \leqslant r < \frac{\alpha}{8\beta}$, *then under* $m = p^\alpha$ *we have the equality*

$$N_m^{(P)}(\delta_1 \ldots \delta_n) = \frac{1}{q_n} P + O\left(rP^{1-\frac{\gamma}{r^2}}\right),$$

where $\gamma = \frac{1}{2\cdot 10^6}$.

Proof. We determine integers t, b, and h by the equalities (322):

$$0.\delta_1 \ldots \delta_n = \frac{t}{q^n}, \qquad b = \left[\frac{tm}{q^n}\right], \qquad h = \left[\frac{(t+1)m}{q^n}\right] - \left[\frac{tm}{q^n}\right].$$

Then we obtain from (323) and (324)

$$N_m^{(P)}(\delta_1 \ldots \delta_n) = \frac{h}{m} P + R, \tag{338}$$

where by (325) the estimate

$$|R| \leqslant \sum_{z=1}^{p^\alpha - 1} \frac{1}{z} \left| \sum_{x=0}^{P-1} e^{2\pi i \frac{azq^x}{p^\alpha}} \right|$$

holds. Hence it follows that

$$|R| \leqslant \sum_{\nu=0}^{\alpha-1} \sum_{\substack{(z,p^\alpha)=p^\nu, \\ 1\leqslant z<p^\alpha}} \frac{1}{z} \left| \sum_{x=0}^{P-1} e^{2\pi i \frac{azq^x}{p^\alpha}} \right|$$

$$= \sum_{\nu=0}^{\alpha-1} \frac{1}{p^\nu} \sum_{\substack{(z_1,p)=1, \\ 1\leqslant z_1<p^{\alpha-\nu}}} \frac{1}{z_1} \left| \sum_{x=0}^{P-1} e^{2\pi i \frac{az_1q^x}{p^{\alpha-\nu}}} \right|.$$

Since by the assumption $P^r = p^\alpha$ and $3 \leqslant r < \frac{\alpha}{8\beta}$, then choosing $r_\nu = r\frac{\alpha-\nu}{\alpha}$ ($\nu = 1, 2, \ldots, \alpha-1$), we get under $\nu \leqslant \frac{\alpha}{3}$

$$P^{r_\nu} = p^{\alpha-\nu}, \qquad 2 \leqslant r_\nu < \frac{\alpha-\nu}{8\beta}$$

and by Theorem 32 the estimate

$$\left| \sum_{x=0}^{P-1} e^{2\pi i \frac{az_1q^x}{p^{\alpha-\nu}}} \right| < 3\,P^{1-\frac{\gamma}{r_\nu^2}} \leqslant 3P^{1-\frac{\gamma}{r^2}} \qquad \left(\gamma = \frac{1}{2\cdot 10^6}\right)$$

holds. Using this estimate and applying the trivial estimation under $\nu \geqslant \frac{\alpha}{3}$ we obtain

$$|R| \leqslant 3\,P^{1-\frac{\gamma}{r^2}} \sum_{0\leqslant\nu<\frac{\alpha}{3}} \frac{1}{p^\nu} \sum_{z_1=1}^{p^{\alpha-\nu}} \frac{1}{z_1} + P \sum_{\frac{\alpha}{3}\leqslant\nu<\alpha} \frac{1}{p^\nu} \sum_{z_1=1}^{p^{\alpha-\nu}} \frac{1}{z_1}$$

$$\leqslant \frac{p}{p-1}\left(3P^{1-\frac{\gamma}{r^2}} + P\,p^{-\frac{\alpha}{3}}\right)(1+\log p^\alpha) = O\left(r\,P^{1-\frac{\gamma}{r^2}} \log P\right).$$

Now, observing that

$$h = \frac{m}{q^n} + O(1),$$

we get the theorem assertion from (338)

$$N_m^{(P)}(\delta_1 \ldots \delta_n) = \frac{1}{q^n}\,P + R + O(1) = \frac{1}{q^n}\,P + O\left(r\,P^{1-\frac{\gamma}{r^2}} \log P\right).$$

§ 23. Connection between exponential sums, quadrature formulas and fractional parts distribution

As it was noted in the introduction, there exists a close connection between estimates of exponential sums and approximate calculation of multiple integrals

$$\int_0^1 \cdots \int_0^1 f(x_1, \ldots, x_s)\, dx_1 \ldots dx_s.$$

This connection is established especially simply, if the function $f(x_1, \ldots, x_s)$ has period 1 with respect to every variable $x_1, \ldots, x_s$ and the Fourier expansion

$$f(x_1, \ldots, x_s) = \sum_{m_1, \ldots, m_s = -\infty}^{\infty} C(m_1, \ldots, m_s) e^{2\pi i (m_1 x_1 + \ldots + m_s x_s)} \tag{339}$$

converges absolutely.

Consider a quadrature formula

$$\int_0^1 \cdots \int_0^1 f(x_1, \ldots, x_s)\, dx_1 \ldots dx_s = \frac{1}{P} \sum_{k=1}^{P} f\big(\xi_1(k), \ldots, \xi_s(k)\big) - R_P[f],$$

where $-R_P[f]$ stands for the error obtained in replacing the integral by the arithmetic mean of the integrand values calculated at the points

$$M_k = M\big(\xi_1(k), \ldots, \xi_s(k)\big) \qquad (k = 1, 2, \ldots, P).$$

The set of points M_k is called a *net*, and the points are said to be *nodes* of the quadrature formula.

Let a certain system of uniformly distributed functions $f_1(x), \ldots, f_s(x)$ be given. Then under any choice of quantities $\gamma_\nu \in (0, 1]$ $(\nu = 1, 2, \ldots, s)$ the number of fulfilments of the inequalities

$$0 \leqslant \{f_1(k)\} < \gamma_1, \ \ldots, \ 0 \leqslant \{f_s(k)\} < \gamma_s \qquad (k = 1, 2, \ldots, P) \tag{340}$$

is equal to $\gamma_1 \ldots \gamma_s P + o(P)$. If coordinates of the quadrature formula nodes are determined by the equalities

$$\xi_1(k) = \{f_1(k)\}, \ \ldots, \ \xi_s(k) = \{f_s(k)\} \qquad (k = 1, 2, \ldots, P),$$

then the nodes are uniformly distributed in the s-dimensional unit cube. In this case by the Weyl criterion the equality

$$\sum_{k=1}^{P} e^{2\pi i (m_1 \xi_1(k) + \ldots + m_s \xi_s(k))} = o(P) \tag{341}$$

holds under any choice of integers $m_1, \dots, m_s$ not all zero. We shall denote the sums (341) by $S(m_1, \dots, m_s)$:

$$S(m_1, \dots, m_s) = \sum_{k=1}^{P} e^{2\pi i (m_1 \xi_1(k) + \dots + m_s \xi_s(k))}$$

and call *exponential sums corresponding to the net of the quadrature formula.*

THEOREM 35. *Let the Fourier series of a function $f(x_1, \dots, x_s)$ converge absolutely, $C(m_1, \dots, m_s)$ be its Fourier coefficients and $S(m_1, \dots, m_s)$ be exponential sums corresponding to the net of a quadrature formula*

$$\int_0^1 \cdots \int_0^1 f(x_1, \dots, x_s)\, dx_1 \dots dx_s = \frac{1}{P} \sum_{k=1}^{P} f(\xi_1(k), \dots, \xi_s(k)) - R_P[f].$$

*Then the equality**

$$R_P[f] = \frac{1}{P} \sum_{m_1, \dots, m_s = -\infty}^{\infty}{}' C(m_1, \dots, m_s) S(m_1, \dots, m_s) \qquad (342)$$

holds and the error $R_P[f]$ tends to zero as $P \to \infty$, if and only if the nodes of the quadrature formula are uniformly distributed in the s-dimensional unit cube.

Proof. Since

$$C(0, \dots, 0) = \int_0^1 \cdots \int_0^1 f(x_1, \dots, x_s)\, dx_1 \dots dx_s,$$

then using the expansion of $f(x_1, \dots, x_s)$ in the Fourier series we get

$$R_P[f] = \frac{1}{P} \sum_{k=1}^{P} f(\xi_1(k), \dots, \xi_s(k)) - \int_0^1 \cdots \int_0^1 f(x_1, \dots, x_s)\, dx_1 \dots dx_s$$

$$= \frac{1}{P} \sum_{k=1}^{P} \sum_{m_1, \dots, m_s = -\infty}^{\infty} C(m_1, \dots, m_s) e^{2\pi i (m_1 \xi_1(k) + \dots + m_s \xi_s(k))} - C(0, \dots, 0).$$

Hence after singling out the summand with $(m_1, \dots, m_s) = (0, \dots, 0)$ and changing the order of summation we have the equality

$$R_P[f] = \frac{1}{P} \sum_{m_1, \dots, m_s = -\infty}^{\infty}{}' C(m_1, \dots, m_s) \sum_{k=1}^{P} e^{2\pi i (m_1 \xi_1(k) + \dots + m_s \xi_s(k))},$$

*Henceforward $\sum'$ signifies that the summation is over s-tuples $(m_1, \dots, m_s) \neq (0, \dots, 0)$.

which coincides with the first assertion of the theorem by the definition of the sums $S(m_1,\dots,m_s)$.

Now we turn to the proof of the second assertion. Let the nodes of the quadrature formula be uniformly distributed in the s-dimensional unit cube. Then by (341)

$$S(m_1,\dots,m_s)=\sum_{k=1}^{P} e^{2\pi i(m_1\xi_1(k)+\dots+m_s\xi_s(k))}=o(P). \tag{343}$$

Take an arbitrary $\varepsilon>0$ and choose $m_0=m_0(\varepsilon)$ and $P_0=P_0(\varepsilon)$ so that the estimates

$$\sum\nolimits_1=\sum_{\max|m_\nu|>m_0}|C(m_1,\dots,m_s)|\,|S(m_1,\dots,m_s)|<\frac{\varepsilon}{2}P,$$

$$\sum\nolimits_2=\sum_{\max|m_\nu|\leqslant m_0}{}'\,|C(m_1,\dots,m_s)|\,|S(m_1,\dots,m_s)|<\frac{\varepsilon}{2}P \qquad (P>P_0)$$

hold. (We obtain the first of these estimates using absolute convergence of the Fourier series and the trivial estimate $|S(m_1,\dots,m_s)|\leqslant P$, the second estimate is satisfied by (343).) Using the equality (342), we get

$$\begin{aligned}|R_P[f]|&\leqslant\frac{1}{P}\sum_{m_1,\dots,m_s=-\infty}^{\infty}{}'\,|C(m_1,\dots,m_s)|\,|S(m_1,\dots,m_s)|\\&\leqslant\frac{1}{P}\sum\nolimits_1+\frac{1}{P}\sum\nolimits_2<\varepsilon,\end{aligned}$$

and, therefore, $\lim_{P\to\infty}R_P[f]=0$.

Now let it be known that the error of the quadrature formula approaches zero as $P\to\infty$. We choose arbitrary integers $m_1,\dots,m_s$ not all zero and consider the function

$$f(x_1,\dots,x_s)=e^{2\pi i(m_1x_1+\dots+m_sx_s)}.$$

Since all Fourier coefficients of this function with the exception of $C(m_1,\dots,m_s)$ vanish and $C(m_1,\dots,m_s)=1$ then by (342)

$$\begin{aligned}|R_P[f]|&=\frac{1}{P}\sum_{n_1,\dots,n_s=-\infty}^{\infty}{}'\,C(n_1,\dots,n_s)S(n_1,\dots,n_s)\\&=\frac{1}{P}S(m_1,\dots,m_s)=\frac{1}{P}\sum_{k=1}^{P}e^{2\pi i(m_1\xi_1(k)+\dots+m_s\xi_s(k))}.\end{aligned}$$

Therefore,

$$\lim_{P\to\infty}\frac{1}{P}\sum_{k=1}^{P}e^{2\pi i(m_1\xi_1(k)+\dots+m_s\xi_s(k))}=\lim_{P\to\infty}\frac{1}{P}R_P[f]=0, \tag{344}$$

and the system of functions $\xi_1(k), \ldots, \xi_s(k)$ is uniformly distributed in the s-dimensional unit cube by the Weyl criterion. Thus the nodes of the quadrature formula are uniformly distributed. The theorem is proved completely.

Let $\alpha > 1$ be an arbitrary real number, m be an integer, $\overline{m} = \max(1, |m|)$ and C be a certain positive constant. We shall say that a function $f(x_1, \ldots, x_s)$ belongs to the class $E_s^\alpha(C)$, if its Fourier coefficients satisfy the condition

$$\left|C(m_1, \ldots, m_s)\right| \leqslant \frac{C}{(\overline{m}_1 \ldots \overline{m}_s)^\alpha}. \tag{345}$$

Using Theorem 35, it is easy to estimate the error of numerical integration of functions $f(x_1, \ldots, x_s)$ belonging to the class $E_s^\alpha(C)$.

Indeed, by (342)

$$\left|R_P[f]\right| \leqslant \frac{1}{P} \sum_{m_1, \ldots, m_s = -\infty}^{\infty}{}' \left|C(m_1, \ldots, m_s)\right| \left|S(m_1, \ldots, m_s)\right| \tag{346}$$

and therefore by (345)

$$\left|R_P[f]\right| \leqslant \frac{C}{P} \sum_{m_1, \ldots, m_s = -\infty}^{\infty}{}' \frac{\left|S(m_1, \ldots, m_s)\right|}{(\overline{m}_1 \ldots \overline{m}_s)^\alpha}. \tag{347}$$

Obviously, the estimate (347) becomes better, when the absolute values of the exponential sums $S(m_1, \ldots, m_s)$ are lesser (especially under small magnitude of the products $\overline{m}_1 \ldots \overline{m}_s$). The collection of the values of these sums depends on the net $M_k = M(\xi_1(k), \ldots, \xi_s(k))$ $(k = 1, 2, \ldots, P)$ only. Therefore picking out such a net that the sums $S(m_1, \ldots, m_s)$ have sufficiently good estimates, it is possible to have influence on a degree of precision of corresponding quadrature formulas.

Let us show that the estimate (347) cannot be improved for functions $f(x_1, \ldots, x_s)$ belonging to the class $E_s^\alpha(C)$. Indeed, let quantities $C_0(m_1, \ldots, m_s)$ be given with the aid of the equalities

$$C_0(m_1, \ldots, m_s) = \begin{cases} \dfrac{C}{(\overline{m}_1 \ldots \overline{m}_s)^\alpha} \dfrac{\left|S(m_1, \ldots, m_s)\right|}{S(m_1, \ldots, m_s)} & \text{if} \quad S(m_1, \ldots, m_s) \neq 0, \\ \dfrac{C}{(\overline{m}_1 \ldots \overline{m}_s)^\alpha} & \text{if} \quad S(m_1, \ldots, m_s) = 0 \end{cases}$$

and a function $f_0(x_1, \ldots, x_s)$ be determined by its Fourier expansion:

$$f_0(x_1, \ldots, x_s) = \sum_{m_1, \ldots, m_s = -\infty}^{\infty} C_0(m_1, \ldots, m_s) e^{2\pi i (m_1 x_1 + \ldots + m_s x_s)}.$$

Since, evidently, the estimate

$$\left|C_0(m_1,\ldots,m_s)\right| \leqslant \frac{C}{(\overline{m}_1\ldots\overline{m}_s)^\alpha}$$

holds, the function $f_0(x_1,\ldots,x_s)$ belongs to the class $E_s^\alpha(C)$. By Theorem 35 we obtain

$$PR_P[f_0] = \sum_{m_1,\ldots,m_s=-\infty}^{\infty}{}' C_0(m_1,\ldots,m_s)S(m_1,\ldots,m_s)$$
$$= \sum_{m_1,\ldots,m_s=-\infty}^{\infty}{}'' C_0(m_1,\ldots,m_s)S(m_1,\ldots,m_s),$$

where in the sum $\sum''$ not only the s-tuple $(0,\ldots,0)$ but s-tuples $(m_1,\ldots,m_s)$, for which $S(m_1,\ldots,m_s) = 0$, are excluded from the range of summation. But then using the determination of the quantities $C_0(m_1,\ldots,m_s)$ we get

$$R_P[f_0] = \frac{C}{P}\sum_{m_1,\ldots,m_s=-\infty}^{\infty}{}'' \frac{|S(m_1,\ldots,m_s)|}{(\overline{m}_1\ldots\overline{m}_s)^\alpha}$$
$$= \frac{C}{P}\sum_{m_1,\ldots,m_s=-\infty}^{\infty}{}' \frac{|S(m_1,\ldots,m_s)|}{(\overline{m}_1\ldots\overline{m}_s)^\alpha},$$

and, therefore, the estimate (347) cannot be improved.

It follows from Theorem 35 that the relation

$$\lim_{P\to\infty}\frac{1}{P}\sum_{k=1}^{P} f(\xi_1(k),\ldots,\xi_s(k)) = \int_0^1\cdots\int_0^1 f(x_1,\ldots,x_s)\,dx_1\ldots dx_s$$

is satisfied if and only if the points $M(\xi_1(k),\ldots,\xi_s(k))$ are uniformly distributed in the s-dimensional unit cube. Note that this equality (see [49]) is valid not only for the functions $f(x_1,\ldots,x_s)$, whose Fourier series are absolutely convergent (as it was shown in Theorem 35), but for arbitrary Riemann-integrable functions as well.

Let a system of functions $f_1(x),\ldots,f_s(x)$ be uniformly distributed in the s-dimensional unit cube and $\gamma_1,\ldots,\gamma_s$ be arbitrary numbers from the interval $(0,1]$. As in § 20, we denote by $N_P(\gamma_1,\ldots,\gamma_s)$ the number of points

$$M_k = M\big(\{f_1(k)\},\ldots,\{f_s(k)\}\big) \qquad (k = 1,2,\ldots,P)$$

falling into the region $0 \leqslant x_1 < \gamma_1$, ..., $0 \leqslant x_s < \gamma_s$ and determine $R_P(\gamma_1,\ldots,\gamma_s)$ with the help of the equality

$$R_P(\gamma_1,\ldots,\gamma_s) = \frac{1}{P}N_P(\gamma_1,\ldots,\gamma_s) - \gamma_1\ldots\gamma_s. \tag{348}$$

Different characteristics of a degree of uniformity of distribution of points M_k in the s-dimensional unit cube are considered in the theory of uniform distribution. The discrepancy $D(P)$ and the mean square discrepancy $T(P)$ defined by the equalities

$$D(P) = \sup_{\gamma_1,\ldots,\gamma_s} |R_P(\gamma_1,\ldots,\gamma_s)|$$

and

$$T^2(P) = \int_0^1 \ldots \int_0^1 R_P^2(x_1,\ldots,x_s)\, dx_1 \ldots dx_s,$$

respectively, belong to such characteristics.

A relation which enables us to estimate the error of quadratic formulas via the above characteristics of uniformity of distribution of net points will be established in the following theorem. We shall say that a function $f(x_1,\ldots,x_s)$ belongs to the class $W_s(C)$, if the conditions

$$\begin{gathered} f(x_1,\ldots,x_{\nu-1},1,x_{\nu+1},\ldots,x_s) = 0 \qquad (\nu = 1,2,\ldots,s), \\ \int_0^1 \cdots \int_0^1 \left(\frac{\partial^s f(x_1,\ldots,x_s)}{\partial x_1 \ldots \partial x_s}\right)^2 dx_1 \ldots dx_s \leqslant C \end{gathered} \tag{349}$$

are satisfied and its partial derivatives

$$\frac{\partial^n f(x_1,\ldots,x_s)}{\partial x_1^{n_1} \ldots \partial x_s^{n_s}} \qquad (0 \leqslant n \leqslant s,\ 0 \leqslant n_1,\ldots,n_s \leqslant 1)$$

are continuous with respect to variables with $n_j = 0$ and satisfy the Dirichlet conditions with respect to other variables.

THEOREM 36. *Let $f(x_1,\ldots,x_s)$ be an arbitrary function from the class $W_s(C)$ and the quantity $R_P(\gamma_1,\ldots,\gamma_s)$ be determined by the equality (348) constituted for coordinates of the net of quadrature formula*

$$\int_0^1 \cdots \int_0^1 f(x_1,\ldots,x_s)dx_1 \ldots dx_s = \frac{1}{P}\sum_{k=1}^{P} f(\xi_1(k),\ldots,\xi_s(k)) - R_P[f]. \tag{350}$$

Then for the error of the formula (350) we have the relations

$$R_P[f] = (-1)^s \int_0^1 \cdots \int_0^1 \frac{\partial^s f(x_1,\ldots,x_s)}{\partial x_1 \ldots \partial x_s} R_P(x_1,\ldots,x_s)\, dx_1 \ldots dx_s,$$

$$\big|R_P[f]\big| \leqslant \sqrt{C}\, T(P),$$

where $T(P)$ is the mean square discrepancy of the net.

Proof. Using the first of the conditions (349), we obtain

$$\frac{\partial^{\nu-1} f(x_1,\dots,x_{\nu-1},1,x_{\nu+1},\dots,x_s)}{\partial x_1\dots\partial x_{\nu-1}} = 0 \qquad (\nu = 1,2,\dots,s).$$

But then, obviously,

$$\int\limits_{\xi_\nu(k)}^{1} \frac{\partial^{\nu} f(x_1,\dots,x_\nu,\xi_{\nu+1}(k),\dots,\xi_s(k))}{\partial x_1\dots\partial x_\nu}\, dx_\nu$$

$$= \frac{\partial^{\nu-1} f(x_1,\dots,x_\nu,\xi_{\nu+1}(k),\dots,\xi_s(k))}{\partial x_1\dots\partial x_{\nu-1}}\Bigg|_{x_\nu=\xi_\nu(k)}^{x_\nu=1}$$

$$= -\frac{\partial^{\nu-1} f(x_1,\dots,x_{\nu-1},\xi_\nu(k),\dots,\xi_s(k))}{\partial x_1\dots\partial x_{\nu-1}},$$

and, therefore,

$$\int\limits_{\xi_1(k)}^{1}\cdots\int\limits_{\xi_s(k)}^{1} \frac{\partial^{s} f(x_1,\dots,x_s)}{\partial x_1\dots\partial x_s}\, dx_1\dots dx_s$$

$$= -\int\limits_{\xi_1(k)}^{1}\cdots\int\limits_{\xi_{s-1}(k)}^{1} \frac{\partial^{s-1} f(x_1,\dots,x_{s-1},\xi_s(k))}{\partial x_1\dots\partial x_{s-1}}\, dx_1\dots dx_{s-1} = \dots$$

$$= (-1)^s f\big(\xi_1(k),\dots,\xi_s(k)\big). \tag{351}$$

Further observing that

$$\int\limits_{0}^{1} \frac{\partial^{\nu} f(x_1,\dots,x_s)}{\partial x_1\dots\partial x_\nu}\, x_\nu\, dx_\nu = x_\nu \frac{\partial^{\nu-1} f(x_1,\dots,x_s)}{\partial x_1\dots\partial x_{\nu-1}}\Bigg|_0^1 - \int\limits_0^1 \frac{\partial^{\nu-1} f(x_1,\dots,x_s)}{\partial x_1\dots\partial x_{\nu-1}}\, dx_\nu$$

$$= -\int\limits_0^1 \frac{\partial^{\nu-1} f(x_1,\dots,x_s)}{\partial x_1\dots\partial x_{\nu-1}}\, dx_\nu,$$

we get

$$\int\limits_0^1\cdots\int\limits_0^1 \frac{\partial^{s} f(x_1,\dots,x_s)}{\partial x_1\dots\partial x_s}\, x_1\dots x_s\, dx_1\dots dx_s$$

$$= -\int\limits_0^1\cdots\int\limits_0^1 \frac{\partial^{s-1} f(x_1,\dots,x_s)}{\partial x_1\dots\partial x_{s-1}}\, x_1\dots x_{s-1}\, dx_1\dots dx_s = \dots$$

$$= (-1)^s \int\limits_0^1\cdots\int\limits_0^1 f(x_1,\dots,x_s)\, dx_1\dots dx_s. \tag{352}$$

We determine a function $\psi(x, y)$ with the aid of the equality

$$\psi(x,y) = \begin{cases} 1 & \text{if } \ x < \gamma, \\ 0 & \text{if } \ x \geqslant \gamma. \end{cases} \tag{353}$$

Then for the number of the net points $M(\xi_1(k), \ldots, \xi_s(k))$ $(k = 1, 2, \ldots, P)$ lying in the region $0 \leqslant x_1 < \gamma_1, \ldots, 0 \leqslant x_s < \gamma_s$ we get

$$N_P(\gamma_1, \ldots, \gamma_s) = \sum_{k=1}^{P} \psi(\xi_1(k), \gamma_1) \ldots \psi(\xi_s(k), \gamma_s)$$

and by (348)

$$R_P(\gamma_1, \ldots, \gamma_s) = \frac{1}{P} \sum_{k=1}^{P} \psi(\xi_1(k), \gamma_1) \ldots \psi(\xi_s(k), \gamma_s) - \gamma_1 \ldots \gamma_s. \tag{354}$$

Using the equalities (351) and (352), we write down the error of the quadrature formula in the form

$$\begin{aligned} R_P[f] &= \frac{1}{P} \sum_{k=1}^{P} f(\xi_1(k), \ldots, \xi_s(k)) - \int_0^1 \cdots \int_0^1 f(x_1, \ldots, x_s)\, dx_1 \ldots dx_s \\ &= (-1)^s \left(\frac{1}{P} \sum_{k=1}^{P} \int_{\xi_1(k)}^{1} \cdots \int_{\xi_s(k)}^{1} \frac{\partial^s f(x_1, \ldots, x_s)}{\partial x_1 \ldots \partial x_s}\, dx_1 \ldots dx_s \right. \\ &\quad \left. - \int_0^1 \cdots \int_0^1 \frac{\partial^s f(x_1, \ldots, x_s)}{\partial x_1 \ldots \partial x_s}\, x_1 \ldots x_s\, dx_1 \ldots dx_s \right). \end{aligned}$$

Hence observing that

$$\begin{aligned} &\int_{\xi_1(k)}^{1} \cdots \int_{\xi_s(k)}^{1} \frac{\partial^s f(x_1, \ldots, x_s)}{\partial x_1 \ldots \partial x_s}\, dx_1 \ldots dx_s \\ &\qquad = \int_0^1 \cdots \int_0^1 \frac{\partial^s f(x_1, \ldots, x_s)}{\partial x_1 \ldots \partial x_s} \prod_{\nu=1}^{s} \psi(\xi_\nu(k), x_\nu)\, dx_1 \ldots dx_s, \end{aligned}$$

by virtue of (354) we obtain the first assertion of the theorem:

$$\begin{aligned} &R_P[f] \\ &= (-1)^s \int_0^1 \cdots \int_0^1 \frac{\partial^s f(x_1, \ldots, x_s)}{\partial x_1 \ldots \partial x_s} \left(\frac{1}{P} \sum_{k=1}^{P} \prod_{\nu=1}^{s} \psi(\xi_\nu(k), x_\nu) - x_1 \ldots x_s \right) dx_1 \ldots dx_s \\ &= (-1)^s \int_0^1 \cdots \int_0^1 \frac{\partial^s f(x_1, \ldots, x_s)}{\partial x_1 \ldots \partial x_s} R_P(x_1, \ldots, x_s)\, dx_1 \ldots dx_s. \end{aligned} \tag{355}$$

The second assertion of the theorem follows from this equality immediately:

$$R_P^2[f] = \left(\int_0^1 \cdots \int_0^1 \frac{\partial^s f(x_1, \ldots, x_s)}{\partial x_1 \ldots \partial x_s} R_P(x_1, \ldots, x_s)\, dx_1 \ldots dx_s \right)^2$$
$$\leqslant \int_0^1 \cdots \int_0^1 \left(\frac{\partial^s f(x_1, \ldots, x_s)}{\partial x_1 \ldots \partial x_s} \right)^2 dx_1 \ldots dx_s$$
$$\times \int_0^1 \cdots \int_0^1 R_P^2(x_1, \ldots, x_s)\, dx_1 \ldots dx_s \leqslant C T^2(P),$$

$$\left| R_P[f] \right| \leqslant \sqrt{C}\, T(P). \tag{356}$$

Note. For the error of approximate integration of functions $f \in W_s(C)$ we have also the estimate

$$\left| R_P[f] \right| \leqslant \sqrt{C}\, D(P), \tag{357}$$

where $D(P)$ is the discrepancy of the net of the quadrature formula (350).

Indeed, by definition

$$T^2(P) = \int_0^1 \cdots \int_0^1 R_P^2(x_1, \ldots, x_s)\, dx_1 \ldots dx_s$$
$$\leqslant \sup_{x_1, \ldots, x_s} R_P^2(x_1, \ldots, x_s) = D^2(P).$$

Therefore, $T(P) \leqslant D(P)$ and the estimate (357) follows from the estimate (356).

Now we shall show that the estimate

$$\left| R_P[f] \right| \leqslant \sqrt{C}\, T(P)$$

obtained in Theorem 36 cannot be improved.

Indeed, let us determine a function $f_0(x_1, \ldots, x_s)$ with the aid of the equality

$$f_0(x_1, \ldots, x_s) = \frac{\sqrt{C}}{T(P)} \int_{x_1}^1 \cdots \int_{x_s}^1 R_P(y_1, \ldots, y_s)\, dy_1 \ldots dy_s. \tag{358}$$

Evidently, this function satisfies the first of the conditions (349):

$$f_0(x_1, \ldots, x_{\nu-1}, 1, x_{\nu+1}, \ldots, x_s) = 0 \qquad (\nu = 1, 2, \ldots, s).$$

Further it is plain that

$$\frac{\partial^s f_0(x_1, \ldots, x_s)}{\partial x_1 \ldots x_s} = \frac{(-1)^s}{T(P)} \sqrt{C} R_P(x_1, \ldots, x_s).$$

But then

$$\int_0^1 \cdots \int_0^1 \left(\frac{\partial^s f_0(x_1, \ldots, x_s)}{\partial x_1 \ldots \partial x_s} \right)^2 dx_1 \ldots dx_s$$

$$= \frac{C}{T^2(P)} \int_0^1 \cdots \int_0^1 R_P^2(x_1, \ldots, x_s)\, dx_1 \ldots dx_s = C,$$

so the second of the conditions (349) is satisfied also and, therefore, $f_0(x_1, \ldots, x_s) \in W_s(C)$. Now applying the equality (355) we obtain

$$R_P[f_0] = (-1)^{-s} \int_0^1 \cdots \int_0^1 \frac{\partial^s f_0(x_1, \ldots, x_s)}{\partial x_1 \ldots \partial x_s} R_P(x_1, \ldots, x_s)\, dx_1 \ldots dx_s$$

$$= \frac{\sqrt{C}}{T(P)} \int_0^1 \cdots \int_0^1 R_P^2(x_1, \ldots, x_s)\, dx_1 \ldots dx_s = \sqrt{C}\, T(P),$$

i.e., the error of the approximate integration of the function $f_0(x_1, \ldots, x_s)$ is equal to $\sqrt{C}\, T(P)$ and the estimate (356) cannot be strengthened.

The obtained results enable us to compare the quality of quadrature formulas nets. In fact, let us consider quadrature formulas based on nets $M_k^{(1)}$ and $M_k^{(2)}$ $(k = 1, 2, \ldots, P)$. Denote by $T_1(P)$ and $T_2(P)$ mean square discrepancies of these nets and suppose that $T_1(P) < T_2(P)$. If a function $f(x_1, \ldots, x_s)$ belongs to the class $W_s(C)$, then by Theorem 36 the error of the first of the quadrature formulas does not exceed $\sqrt{C}\, T_1(P)$. On the other hand, determining $f_0(x_1, \ldots, x_s)$ by the equality (358), we obtain that in the class $W_s(C)$ there is a function for which the error of the second formula $R_P^{(2)}[f_0]$ is greater than $\sqrt{C}\, T_1(P)$:

$$R_P^{(2)}[f_0] = \sqrt{C}\, T_2(P) > \sqrt{C}\, T_1(P).$$

Thus, the following criterion of the quality of quadrature formulas nets is valid: in numerical integration of functions belonging to the class $W_s(C)$, the net with smaller mean square discrepancy is the best of two given nets.

From a definition of the function $\psi(x, y)$ (353), we have the equalities

$$\int_0^1 \psi(x, \gamma)\, \gamma\, d\gamma = \int_x^1 \gamma\, d\gamma = \frac{1 - x^2}{2},$$

$$\int_0^1 \psi(x, \gamma)\, \psi(y, \gamma)\, d\gamma = \int_{\max(x,y)}^1 d\gamma = 1 - \max(x, y).$$

A simple expression to calculate mean square discrepancy can be obtained with the aid of these equalities.

In fact, by (354)

$$R_P^2(\gamma_1,\ldots,\gamma_s) = \left(\frac{1}{P}\sum_{k=1}^{P}\psi(\xi_1(k),\gamma_1)\ldots\psi(\xi_s(k),\gamma_s) - \gamma_1\ldots\gamma_s\right)^2$$
$$= \frac{1}{P^2}\sum_{j,k=1}^{P}\prod_{\nu=1}^{s}\psi(\xi_\nu(j),\,\gamma_\nu)\psi(\xi_\nu(k),\,\gamma_\nu)$$
$$-\frac{2}{P}\sum_{k=1}^{P}\prod_{\nu=1}^{s}\psi(\xi_\nu(k),\,\gamma_\nu)\gamma_\nu + \gamma_1^2\ldots\gamma_s^2,$$

and, therefore,

$$T^2(P) = \int_0^1\ldots\int_0^1 R_P^2(\gamma_1,\ldots,\gamma_s)\,d\gamma_1\ldots d\gamma_s$$
$$= \frac{1}{P^2}\sum_{j,k=1}^{P}\prod_{\nu=1}^{s}\Big[1-\max\big(\xi_\nu(j),\,\xi_\nu(k)\big)\Big]$$
$$-\frac{2}{P}\sum_{k=1}^{P}\prod_{\nu=1}^{s}\frac{1-\xi_\nu^2(k)}{2} + \frac{1}{3^s}. \tag{359}$$

Let, as above, $D(P)$ be the discrepancy of a net M_k $(k = 1,2,\ldots,P)$. It is seen from the estimate

$$\big|R_P[f]\big| \leqslant \sqrt{C}\,D(P), \tag{360}$$

indicated in the note of Theorem 36, that it is possible to judge the quality of quadrature formulas nets by the magnitude of the discrepancy $D(P)$. But there is no explicit expression, similar to (359), for the calculation of the discrepancy; this circumstance prevents practical use of the estimate (360).

The estimate

$$\big|R_P[f]\big| \leqslant \frac{C}{P}\sum_{m_1,\ldots,m_s=-\infty}^{\infty}{}' \frac{|S(m_1,\ldots,m_s)|}{(\overline{m}_1\ldots\overline{m}_s)^\alpha} \tag{361}$$

following by (347) from Theorem 35 can also not always be used for practical comparison of quadrature formulas nets. But this estimate is unimprovable and in some cases enables us to establish convenient criteria for the quality of nets.

§ 24. Quadrature and interpolation formulas with the number-theoretical nets

Let $s \geqslant 2$, p be a prime greater than s, and $(m_1, \ldots, m_s, p) = 1$. By virtue of Lemma 4 the estimate

$$\left| \sum_{k=1}^{p^2} e^{2\pi i \frac{m_1 k + \ldots + m_s k^s}{p^2}} \right| \leqslant (s-1)p \tag{362}$$

holds. Let us determine coordinates of a net M_k by the equalities

$$\xi_1(k) = \left\{ \frac{k}{p^2} \right\}, \ldots, \xi_s(k) = \left\{ \frac{k^s}{p^2} \right\} \qquad (k = 1, 2, \ldots, P)$$

and consider the exponential sum corresponding to the net M_k:

$$S(m_1, \ldots, m_s) = \sum_{k=1}^{P} e^{2\pi i \frac{m_1 k + \ldots + m_s k^s}{p^2}}.$$

If $P = p^2$, then the sum $S(m_1, \ldots, m_s)$ coincides with the sum (362) and therefore under $(m_1, \ldots, m_s, p) = 1$ the estimate

$$\left| S(m_1, \ldots, m_s) \right| \leqslant (s-1)\, p = (s-1)\sqrt{P} \tag{363}$$

is valid for it. The following theorem is based on the use of this estimate.

THEOREM 37. *If a function $f(x_1, \ldots, x_s)$ belongs to the class $E_s^\alpha(C)$, p is a prime greater than s, and $P = p^2$, then for the error of the quadrature formula*

$$\int_0^1 \cdots \int_0^1 f(x_1, \ldots, x_s)\, dx_1 \ldots dx_s = \frac{1}{P} \sum_{k=1}^{P} f\left(\left\{ \frac{k}{p^2} \right\}, \ldots, \left\{ \frac{k^s}{p^2} \right\} \right) - R_P[f]$$

we have the estimate

$$\left| R_P[f] \right| \leqslant \left(\frac{3\alpha}{\alpha - 1} \right)^s \frac{sC}{\sqrt{P}}. \tag{364}$$

Proof. As it was shown in the preceding section, the estimate

$$\left| R_P[f] \right| \leqslant \frac{C}{P} \sum_{m_1, \ldots, m_s = -\infty}^{\infty}{}' \frac{\left| S(m_1, \ldots, m_s) \right|}{(\overline{m}_1 \ldots \overline{m}_s)^\alpha}$$

holds for $f(x_1,\ldots,x_s) \in E_s^\alpha(C)$. Quantities $m_1,\ldots,m_s$, whose greatest common divisor is a multiple of p, may be represented in the form $n_1p,\ldots,n_sp$. Applying in this case the trivial estimate

$$\left|S(n_1p,\ldots,n_sp)\right| \leqslant p^2 = P$$

and using under $(m_1,\ldots,m_s,p)=1$ the estimate (363), we get

$$\begin{aligned}\left|R_P[f]\right| &\leqslant \frac{C}{P} \sum_{n_1,\ldots,n_s=-\infty}^{\infty}{}' \frac{|S(n_1p,\ldots,n_sp)|}{(\overline{n_1p}\ldots\overline{n_sp})^\alpha} \\ &+ \frac{C}{P} \sum_{(m_1,\ldots,m_s,p)=1} \frac{|S(m_1,\ldots,m_s)|}{(\overline{m}_1\ldots\overline{m}_s)^\alpha} \\ &\leqslant C \sum_{n_1,\ldots,n_s=-\infty}^{\infty}{}' \frac{1}{(\overline{n_1p}\ldots\overline{n_sp})^\alpha} \\ &+ \frac{(s-1)C}{\sqrt{P}} \sum_{(m_1,\ldots,m_s,p)=1} \frac{1}{(\overline{m}_1\ldots\overline{m}_s)^\alpha}\,. \end{aligned} \tag{365}$$

Since it follows from the definition of the quantities $\overline{m}$ that

$$\overline{n_\nu p} = \begin{cases} \geqslant \bar{n}_\nu & \text{under every} \quad n_\nu, \\ = \bar{n}_\nu p & \text{under} \quad n_\nu \neq 0, \end{cases}$$

then the estimate

$$\overline{n_1p}\ldots\overline{n_sp} \geqslant p\bar{n}_1\ldots\bar{n}_s$$

holds for every s-tuple of integers $n_1,\ldots,n_s$ not all zero. But then

$$\begin{aligned}\sum_{n_1,\ldots,n_s=-\infty}^{\infty}{}' \frac{1}{(\overline{n_1p}\ldots\overline{n_sp})^\alpha} &\leqslant \frac{1}{p^\alpha} \sum_{n_1,\ldots,n_s=-\infty}^{\infty}{}' \frac{1}{(\bar{n}_1\ldots\bar{n}_s)^\alpha} \\ &< \frac{1}{p^{\frac{\alpha}{2}}}\left(\sum_{n=-\infty}^{\infty} \frac{1}{\bar{n}^\alpha}\right)^s < \left(\frac{3\alpha}{\alpha-1}\right)^s \frac{1}{\sqrt{P}}\,. \end{aligned} \tag{366}$$

Substituting this estimate into (365), we get the theorem assertion

$$\left|R_P[f]\right| \leqslant \left(\frac{3\alpha}{\alpha-1}\right)^s \frac{C}{\sqrt{P}} + \frac{(s-1)C}{\sqrt{P}}\left(\sum_{m=-\infty}^{\infty} \frac{1}{\overline{m}^\alpha}\right)^s < \left(\frac{3\alpha}{\alpha-1}\right)^s \frac{sC}{\sqrt{P}}\,.$$

Note that the estimate

$$R_P[f] = O\left(\frac{1}{\sqrt{P}}\right)$$

obtained in Theorem 37 holds also for the quadrature formula

$$\int_0^1 \cdots \int_0^1 f(x_1, \ldots, x_s)\, dx_1 \ldots dx_s = \frac{1}{P} \sum_{k=1}^{P} f\left(\left\{\frac{k}{p}\right\}, \ldots, \left\{\frac{k^s}{p}\right\}\right) - R_P[f], \qquad (367)$$

where p is a prime greater than s and $P = p$. In the proof of this statement, deeper results from the theory of exponential sums have to be invoked and the estimate of A. Weil

$$\left| \sum_{k=1}^{p} e^{2\pi i \frac{m_1 k + \ldots + m_s k^s}{p}} \right| \leqslant (s-1)\sqrt{p}$$

be used instead of the estimate (362) (see the note of Theorem 7). In other respects the proof does not differ from the proof of the estimate (364).

The quadrature formulas (364) and (367) guarantee the same order of decrease of the error, as is obtained (with the probability close to unity) for quadrature formulas based on Monte Carlo method. We shall consider quadrature formulas, whose error on the class $E_s^\alpha(C)$ has the higher order of decrease, in the following theorem.

Let $p > 2$, $P \leqslant p$, and a_ν be integers relatively prime to p ($\nu = 1, 2, \ldots, s$). Under $P = p$ nets of the form

$$M_k = M\left(\left\{\frac{a_1 k}{p}\right\}, \ldots, \left\{\frac{a_s k}{p}\right\}\right) \qquad (k = 1, 2, \ldots, P)$$

are called *parallelepipedal nets*. Exponential sums corresponding to parallelepipedal nets

$$S(m_1, \ldots, m_s) = \sum_{k=1}^{p} e^{2\pi i \frac{(a_1 m_1 + \ldots + a_s m_s)k}{p}}$$

are, actually, complete rational exponential sums of the first degree. By Lemma 2 for them the equality

$$\begin{aligned} S(m_1, \ldots, m_s) &= p\,\delta_p(a_1 m_1 + \ldots + a_s m_s) \\ &= \begin{cases} p & \text{if} \quad a_1 m_1 + \ldots + a_s m_s \equiv 0 \pmod{p}, \\ 0 & \text{otherwise} \end{cases} \end{aligned} \qquad (368)$$

holds.

Theorem 38. *Let $f(x_1, \ldots, x_s) \in E_s^\alpha(C)$, $p > s$, $(a_\nu, p) = 1$ ($\nu = 1, 2, \ldots, s$), and $P = p$. For the error of quadrature formulas with parallelepipedal nets*

$$\int_0^1 \cdots \int_0^1 f(x_1, \ldots, x_s)\, dx_1 \ldots dx_s$$

$$= \frac{1}{P} \sum_{k=1}^{P} f\left(\left\{\frac{a_1 k}{p}\right\}, \ldots, \left\{\frac{a_s k}{p}\right\}\right) - R_P[f] \qquad (369)$$

we have the estimate

$$\left|R_P[f]\right| \leqslant C \sum_{m_1,\ldots,m_s=-\infty}^{\infty}{}' \frac{\delta_p(a_1m_1+\ldots+a_sm_s)}{(\overline{m}_1\ldots\overline{m}_s)^\alpha}\,; \tag{370}$$

the net of the formula (369) can be chosen so that under any $\alpha < 1$

$$\left|R_P[f]\right| \leqslant CC_1 \frac{\log^{\alpha s} P}{P^\alpha}\,,$$

where $C_1 = C_1(\alpha, s)$ is a constant only depending on α and s.

Proof. It follows from the definition of the class $E_s^\alpha(C)$, that for the Fourier coefficients of the function $f(x_1,\ldots,x_s)$ the estimate

$$|C(m_1,\ldots,m_s)| \leqslant \frac{C}{(\overline{m}_1\ldots\overline{m}_s)^\alpha}$$

holds. But then by Theorem 35

$$\begin{aligned}\left|R_P[f]\right| &\leqslant \frac{1}{P} \sum_{m_1,\ldots,m_s=-\infty}^{\infty}{}' \left|C(m_1,\ldots,m_s)\right|\left|S(m_1,\ldots,m_s)\right| \\ &\leqslant \frac{C}{P} \sum_{m_1,\ldots,m_s=-\infty}^{\infty}{}' \frac{|S(m_1,\ldots,m_s)|}{(\overline{m}_1\ldots\overline{m}_s)^\alpha}\,.\end{aligned}$$

Hence, using the equality (368), we get the first assertion of the theorem:

$$\begin{aligned}\left|R_P[f]\right| &\leqslant \frac{C}{P} \sum_{m_1,\ldots,m_s=-\infty}^{\infty}{}' \frac{p\delta_p(a_1m_1+\ldots+a_sm_s)}{(\overline{m}_1\ldots\overline{m}_s)^\alpha} \\ &= C \sum_{m_1,\ldots,m_s=-\infty}^{\infty}{}' \frac{\delta_p(a_1m_1+\ldots+a_sm_s)}{(\overline{m}_1\ldots\overline{m}_s)^\alpha}\,.\end{aligned} \tag{371}$$

To prove the second assertion we determine integers p_1 and p_2 with the aid of the equalities $p_1 = \left[\frac{p-1}{2}\right]$, $p_2 = \left[\frac{p}{2}\right]$ and replace m_ν by $n_\nu p + m_\nu$ in (371):

$$\left|R_P[f]\right| \leqslant C \sum_{\substack{n_1,\ldots,n_s=-\infty,\\ -p_1\leqslant m_1,\ldots,m_s\leqslant p_2}}^{\infty}{}' \frac{\delta_p(a_1m_1+\ldots+a_sm_s)}{(\overline{n_1p+m_1}\ldots\overline{n_sp+m_s})^\alpha}\,. \tag{372}$$

Since, obviously, under $m \in [-p_1, p_2]$

$$\sum_{n=-\infty}^{\infty} \frac{1}{(\overline{np+m})^\alpha} \leqslant \frac{1}{\overline{m}^\alpha} + 2\sum_{n=1}^{\infty} \frac{1}{\left(np-\frac{p}{2}\right)^\alpha} < \frac{3\alpha}{\alpha-1}\,\frac{1}{\overline{m}^\alpha}$$

and by (366)

$$\sum_{n_1,\dots,n_s=-\infty}^{\infty}{}' \frac{1}{(\overline{n_1 p}\dots\overline{n_s p})^\alpha} < \left(\frac{3\alpha}{\alpha-1}\right)^s \frac{1}{p^\alpha},$$

then, in (372) singling out the summands with $m_1 = \dots = m_s = 0$, we obtain

$$\begin{aligned} |R_P[f]| &\leqslant \left(\frac{3\alpha}{\alpha-1}\right)^s \frac{C}{p^\alpha} \\ &+ C \sum_{m_1,\dots,m_s=-p_1}^{p_2}{}' \sum_{n_1,\dots,n_s=-\infty}^{\infty} \frac{\delta_p(a_1 m_1 + \dots + a_s m_s)}{(\overline{n_1 p + m_1}\dots\overline{n_s p + m_s})^\alpha} \\ &\leqslant \left(\frac{3\alpha}{\alpha-1}\right)^s C\left(\frac{1}{p^\alpha} + \sum_{m_1,\dots,m_s=-p_1}^{p_2}{}' \frac{\delta_p(a_1 m_1 + \dots + a_s m_s)}{(\overline{m}_1\dots\overline{m}_s)^\alpha}\right). \end{aligned} \tag{373}$$

Denote by $T(z_1,\dots,z_s)$ the sum

$$T(z_1,\dots,z_s) = \sum_{m_1,\dots,m_s=-p_1}^{p_2}{}' \frac{\delta_p(m_1 z_1 + \dots + m_s z_s)}{\overline{m}_1\dots\overline{m}_s},$$

where $z_1,\dots,z_s$ are integers. Let p be a prime and the minimum of the function $T(z_1,\dots,z_s)$ in the domain $1 \leqslant z_\nu < p$ $(\nu = 1,2,\dots,s)$ be attained under $z_1 = a_1, \dots, z_s = a_s$:

$$T(a_1,\dots,a_s) = \min_{1\leqslant z_1,\dots,z_s<p} T(z_1,\dots,z_s). \tag{374}$$

Since under $m_1,\dots,m_s$ not divisible by p simultaneously the estimate

$$\sum_{z_1,\dots,z_s=1}^{p-1} \delta_p(m_1 z_1 + \dots + m_s z_s) \leqslant (p-1)^{s-1} \tag{375}$$

holds, then evidently

$$\begin{aligned} T(a_1,\dots,a_s) &\leqslant \frac{1}{(p-1)^s} \sum_{z_1,\dots,z_s=1}^{p-1} T(z_1,\dots,z_s) \\ &= \frac{1}{(p-1)^s} \sum_{m_1,\dots,m_s=-p_1}^{p_2}{}' \frac{1}{\overline{m}_1\dots\overline{m}_s} \sum_{z_1,\dots,z_s=1}^{p-1} \delta_p(m_1 z_1 + \dots + m_s z_s). \end{aligned}$$

Hence it follows by (375) that

$$\begin{aligned} T(a_1,\dots,a_s) &\leqslant \frac{1}{p-1} \sum_{m_1,\dots,m_s=-p_1}^{p_2}{}' \frac{1}{\overline{m}_1\dots\overline{m}_s} \\ &\leqslant \frac{1}{p-1}\left(1 + 2\sum_{m=1}^{p-1}\frac{1}{m}\right)^s < \frac{2(3+2\log p)^s}{p}. \end{aligned} \tag{376}$$

Let us consider the quadrature formula (369), in which the quantities $a_1, \ldots, a_s$ are chosen according to the condition (374). Since $\alpha > 1$, then

$$\sum_{m_1,\ldots,m_s=-p_1}^{p_2}{}' \frac{\delta_p(a_1m_1+\ldots+a_sm_s)}{(\overline{m}_1\ldots\overline{m}_s)^\alpha} \leqslant \left(\sum_{m_1,\ldots,m_s=-p_1}^{p_2}{}' \frac{\delta_p(a_1m_1+\ldots+a_sm_s)}{\overline{m}_1\ldots\overline{m}_s}\right)^\alpha = T^\alpha(a_1,\ldots,a_s).$$

Therefore, using the estimate (376), we obtain

$$\sum_{m_1,\ldots,m_s=-p_1}^{p_2}{}' \frac{\delta_p(a_1m_1+\ldots+a_sm_s)}{(\overline{m}_1\ldots\overline{m}_s)^\alpha} \leqslant \frac{2^\alpha(3+2\log p)^{\alpha s}}{p^\alpha}.$$

Substituting this estimate into (373), we get the second assertion of the theorem under a certain $C_1 < \left(\frac{30\alpha}{\alpha-1}\right)^{\alpha s}$:

$$|R_P[f]| \leqslant \left(\frac{3\alpha}{\alpha-1}\right)^s C\,\frac{1+2^\alpha(3+2\log p)^{\alpha s}}{p^\alpha} \leqslant CC_1\frac{\log^{\alpha s} p}{p^\alpha} = CC_1\frac{\log^{\alpha s} P}{P^\alpha}.$$

If there exists an infinite sequence of positive integers p such that under certain $C_0 = C_0(s)$, $\beta = \beta(s)$, and $a_1 = a_1(p), \ldots, a_s = a_s(p)$ the estimate

$$\sum_{m_1,\ldots,m_s=-p_1}^{p_2}{}' \frac{\delta_p(a_1m_1+\ldots+a_sm_s)}{\overline{m}_1\ldots\overline{m}_s} \leqslant C_0\frac{\log^\beta p}{p}, \tag{377}$$

where $p_1 = \left[\frac{p-1}{2}\right]$ and $p_2 = \left[\frac{p}{2}\right]$, holds, then for every p belonging to the sequence, the integers $a_1, \ldots, a_s$ are called *optimal coefficients modulo p* and the nets

$$M_k = M\left(\left\{\frac{a_1k}{p}\right\}, \ldots, \left\{\frac{a_sk}{p}\right\}\right) \qquad (k = 1, 2, \ldots, p),$$

corresponding to them, are said to be *optimal parallelepipedal nets.*

It is seen from Theorem 38 that optimal parallelepipedal nets enable us to construct quadrature formulas, for the error of which the estimate

$$R_P[f] = O\left(\frac{\log^\gamma P}{P^\alpha}\right) \qquad (\gamma = \alpha s) \tag{378}$$

holds. It can be shown that for any choice of nets it is impossible to obtain the error term better than

$$R_P[f] = O\left(\frac{\log^{s-1} P}{P^\alpha}\right) \tag{379}$$

on classes $E_s^\alpha(C)$. Thus the estimate (378) is close to the best possible in principal order and only the logarithmic factor can be improved.

Let us note some other characteristic peculiarities of quadrature formulas with parallelepipedal nets. It is seen from the estimate (378) that such quadrature formulas react automatically to the smoothness of the integrand: the smoother the periodic function $f(x_1, \ldots, x_s)$, the more precise results are ensured by the application of one and the same quadrature formula. This property of computational algorithms (see [2]) is called their "*insatiableness*". Thus the quadrature formulas with parallelepipedal nets enjoy the property of insatiableness.

Denote by q the minimal value of the product $\overline{m}_1 \ldots \overline{m}_s$ for nontrivial solutions of the congruence

$$a_1 m_1 + \ldots + a_s m_s \equiv 0 \pmod{p}. \tag{380}$$

Another peculiarity of quadrature formulas with parallelepipedal nets is the fact that they are exact for trigonometric polynomials of the form

$$Q(x_1, \ldots, x_s) = \sum_{\overline{m}_1 \ldots \overline{m}_s < q} C(m_1, \ldots, m_s) e^{2\pi i (m_1 x_1 + \ldots + m_s x_s)}, \tag{381}$$

i.e., under $P = p$ for every trigonometrical polynomial (381) the equality

$$\int_0^1 \cdots \int_0^1 Q(x_1, \ldots, x_s)\, dx_1 \ldots dx_s = \frac{1}{P} \sum_{k=1}^{P} Q\left(\left\{\frac{a_1 k}{p}\right\}, \ldots, \left\{\frac{a_s k}{p}\right\}\right) \tag{382}$$

is fulfilled.

Indeed, let us consider the quadrature formula

$$\int_0^1 \cdots \int_0^1 Q(x_1, \ldots, x_s)\, dx_1 \ldots dx_s$$
$$= \frac{1}{P} \sum_{k=1}^{P} Q\left(\left\{\frac{a_1 k}{p}\right\}, \ldots, \left\{\frac{a_s k}{p}\right\}\right) - R_P[Q]. \tag{383}$$

Since under $\overline{m}_1 \ldots \overline{m}_s \geqslant q$ the Fourier coefficients of the polynomial (381) vanish, then by Theorem 35 the equality

$$R_P[f] = \sum_{\overline{m}_1 \ldots \overline{m}_s < q}{}' C(m_1, \ldots, m_s)\, \delta_p(a_1 m_1 + \ldots + a_s m_s) \tag{384}$$

holds. By the definition of the quantity q in the sum (384) there is no s-tuple $m_1, \ldots, m_s$ satisfying the congruence $a_1 m_1 + \ldots + a_s m_s \equiv 0 \pmod{p}$, and, therefore, every term of this sum is equal to zero. But then $R_P[f] = 0$, and we obtain the equality (382) from (383).

The quantity q determined by (380) will be called *the parameter of a parallelepipedal net.* It is seen from the equalities (381) and (382) that the more the net parameter is, the more the number of trigonometric polynomials for which the corresponding quadrature formula is exact. Therefore in construction of quadrature formulas it is appropriate to use nets with the largest possible values of the parameter q. The optimal parallelepipedal nets are these very nets. To be sure of it, we consider the inequality (377)

$$\sum_{m_1,\dots,m_s=-p_1}^{p_2}{}' \frac{\delta_p(a_1m_1+\ldots+a_sm_s)}{\overline{m}_1\ldots\overline{m}_s} \leqslant C_0\frac{\log^\beta p}{p}$$

used in defining optimal coefficients. Since in the sum $\sum'_{m_1,\dots,m_s}$ there is an s-tuple of values $m_1,\dots,m_s$, which satisfy the relations

$$\overline{m}_1\ldots\overline{m}_s = q, \qquad a_1m_1+\ldots+a_sm_s \equiv 0 \pmod p,$$

then this sum contains a term being equal to $\frac{1}{q}$, and therefore

$$\frac{1}{q} \leqslant \sum_{m_1,\dots,m_s=-p_1}^{p_2}{}' \frac{\delta_p(a_1m_1+\ldots+a_sm_s)}{\overline{m}_1\ldots\overline{m}_s} \leqslant C_0\frac{\log^\beta p}{p}, \qquad q \geqslant \frac{p}{C_0\log^\beta p}.$$

On the other hand, it is seen from the definition (380) that $q \leqslant p$. Thus the parameter of optimal parallelepipedal nets differs from its largest possible value not more than by a certain power of the logarithm.

In connection with the needs of computational practice, there arises a question about economical algorithms for computing optimal coefficients. Under $s = 2$ this question is easily solved with the aid of properties of finite continued fractions. Let $1 < a < p$, $(a,p) = 1$ and partial quotients of the continued fraction expansion of the number $\frac{a}{p}$ be bounded by a certain constant M:

$$\frac{a}{p} = \cfrac{1}{q_1+\cfrac{1}{q_2+\ddots+\cfrac{1}{q_n}}}, \qquad q_\nu \leqslant M \quad (\nu = 1,2,\dots,n). \tag{385}$$

We shall show that the numbers $1, a$ are optimal coefficients modulo p.

Indeed, take $s = 2$, $a_1 = 1$ and $a_2 = a$ in the sum (377). Observing that the summands with $m_1 = 0$ or $m_2 = 0$ vanish and using the equality $\overline{m}_\nu = |m_\nu|$ for $m_\nu \neq 0$, we obtain

$$\sum_{m_1,m_2=-p_1}^{p_2}{}' \frac{\delta_p(m_1+am_2)}{\overline{m}_1\overline{m}_2} = \sum_{m_1=-p_1}^{p_2}{}' \sum_{m_2=-p_1}^{p_2}{}' \frac{\delta_p(m_1+am_2)}{|m_1|\,|m|_2}. \tag{386}$$

Since $|m_1| \leqslant p_2 \leqslant \frac{1}{2}p$, then the quantities m_1, m_2, for which the terms of the sum (386) are not equal to zero, satisfy the relations

$$am_2 \equiv -m_1 \pmod p,$$

$$\left\|\frac{am_2}{p}\right\| = \left\|\frac{-m_1}{p}\right\| = \frac{|m_1|}{p}, \qquad |m_1| = p\left\|\frac{am_2}{p}\right\|.$$

But then it follows from (386) that

$$\sum_{m_1,m_2=-p_1}^{p_2}{}' \frac{\delta_p(m_1+am_2)}{\overline{m}_1\overline{m}_2} = \frac{1}{p}\sum_{m_2=-p_1}^{p_2}{}' \frac{1}{|m_2|\left\|\frac{am_2}{p}\right\|} \leqslant \frac{2}{p}\sum_{m_2=1}^{p_2} \frac{1}{|m_2|\left\|\frac{am_2}{p}\right\|}.$$

Hence using Lemma 3, we get

$$\sum_{m_1,m_2=-p_1}^{p_2}{}' \frac{\delta_p(m_1+am_2)}{\overline{m}_1\overline{m}_2} \leqslant 36M\frac{\log^2 p}{p}$$

and the integers $1, a$ are optimal coefficients modulo p by the definition (377).

In particular, under $M = 1$ all partial quotients of the fraction (385) equal 1, and numerators and denominators of its convergents are successive terms of the Fibonacci sequence

$$1, 1, 2, 3, 5, 8, \ldots, Q_n, \ldots,$$
$$Q_0 = 1, \quad Q_1 = 1, \quad Q_n = Q_{n-1} + Q_{n-2} \qquad (n \geqslant 2).$$

Thus under any $n > 2$ the numbers $1, Q_{n-1}$ are optimal coefficients modulo Q_n. It can be shown that under $a = Q_{n-1}$, $p = Q_n$ and $P = p$ for functions belonging to the class $E_2^\alpha(C)$ the error of the quadrature formula

$$\int_0^1\int_0^1 f(x_1,x_2)\,dx_1dx_2 = \frac{1}{P}\sum_{k=1}^{P} f\left(\left\{\frac{k}{p}\right\}\left\{\frac{ak}{p}\right\}\right) - R_P[f]$$

is estimated especially well:

$$R_P[f] = O\left(\frac{\log P}{P^2}\right).$$

It follows from (379) that the order of this estimate cannot be improved under any choice of nets.

If the multiplicity s of the integral is greater than 2, then algorithms for optimal coefficients computation are more complicated. We shall expose some of them without any details.

Let p be a prime greater than s and $(z,p)=1$. We determine functions $T(z)$ and $H(z)$ by the equalities:

$$T(z)=\frac{1}{p}\sum_{k=1}^{p-1}\left(1-\log 4\sin^2\pi\left\{\frac{zk}{p}\right\}\right)\dots\left(1-\log 4\sin^2\pi\left\{\frac{z^sk}{p}\right\}\right),$$

$$H(z)=\frac{3^s}{p}\sum_{k=1}^{p}\left(1-2\left\{\frac{zk}{p}\right\}\right)^2\dots\left(1-2\left\{\frac{z^sk}{p}\right\}\right)^2.$$

If the minimum of $T(z)$ or $H(z)$ for integers z from the interval $1\leqslant z<p$ is attained at $z=a$, then the least positive residues of numbers $1,a,\dots,a^{s-1}$ are optimal coefficients modulo p. The theorem about the number of solutions of polynomial congruences to a prime modulus and the form of coefficients of the Fourier series for the functions $1-\log 4\sin^2\pi x$ and $3(1-2\{x\})^2$,

$$1-\log 4\sin^2\pi x=1+\sum_{m=-\infty}^{\infty}{}'\,\frac{e^{2\pi i mx}}{|m|}\qquad(\{x\}\neq 0),$$

$$3(1-2\{x\})^2=1+\frac{6}{\pi^2}\sum_{m=-\infty}^{\infty}{}'\,\frac{e^{2\pi i mx}}{m^2},$$

are used to prove this assertion.

The number of elementary arithmetic operations in the minimization of the functions $T(z)$ and $H(z)$ has order $O(p^2)$ and requires long calculations. Nevertheless the table of optimal coefficients for computing integrals, whose multiplicity does not exceed ten, was obtained with the aid of a slight modification of the above algorithms. Recently more economical algorithms are found, the number of operations in them is reduced to $O(p)$. That decreases essentially the volume of preliminary calculations and extends possibilities of approximate computation of integrals by the method of optimal coefficients.

The number-theoretical quadrature formulas can be used in a number of problems of analysis and mathematical physics. Here we restrict ourselves to one example illustrating an approach to the construction of interpolation formulas for functions of several variables.

For the sake of writing convenience under $s>1$ we shall use the notation

$$f^{n_1,\dots,n_s}(x_1,\dots,x_s)=\frac{\partial^{n_1+\dots+n_s}f(x_1,\dots,x_s)}{\partial x_1^{n_1}\dots\partial x_s^{n_s}}.$$

LEMMA 33. *Let $\alpha>2$ and a function $f(x_1,\dots,x_s)$ belongs to the class $E_s^\alpha(C)$. Then we have the equality*

$$f(x_1,\dots,x_s)=\sum_{\tau_1,\dots,\tau_s=0}^{1}\int_0^1\dots\int_0^1 f^{\tau_1,\dots,\tau_s}(y_1,\dots,y_s)$$

$$\times \prod_{\nu=1}^{s} \left(\{y_\nu - x_\nu\} - \frac{1}{2} \right)^{\tau_\nu} dy_1 \dots dy_s. \tag{387}$$

Proof. At first we observe that under $\alpha > 2$ the fact, that the function $f(x_1, \dots, x_s)$ belongs to the class $E_s^\alpha(C)$, implies the existence and continuity of the derivatives

$$f^{\tau_1,\dots,\tau_s}(x_1, \dots, x_s) \qquad (\tau_\nu = 0, 1,\ \nu = 1, 2, \dots, s).$$

Let $s = 1$, $\alpha > 2$, and $f(x) \in E_1^\alpha(C)$. Performing the integration by parts and using the periodicity of the integrands, we obtain

$$\int_0^1 f'(y)\left(\{y - x\} - \frac{1}{2}\right) dy = \int_0^1 f'(x + y)\left(y - \frac{1}{2}\right) dy$$

$$= \left(y - \frac{1}{2}\right) f(x + y)\Big|_0^1 - \int_0^1 f(x + y)\, dy = f(x) - \int_0^1 f(y)\, dy,$$

and, therefore,

$$f(x) = \int_0^1 f(y)\, dy + \int_0^1 f'(y)\left(\{y - x\} - \frac{1}{2}\right) dy = \sum_{\tau=0}^{1} \int_0^1 f^{(\tau)}(y)\left(\{y - x\} - \frac{1}{2}\right)^\tau dy.$$

Applying this equality to the variables $x_1, \dots, x_s$ consecutively, we get the lemma assertion:

$$f(x_1, \dots, x_s) = \sum_{\tau_1=0}^{1} \int_0^1 f^{\tau_1}(y_1, x_2, \dots, x_s)\left(\{y_1 - x_1\} - \frac{1}{2}\right)^{\tau_1} dy_1 = \dots$$

$$= \sum_{\tau_1,\dots,\tau_s=0}^{1} \int_0^1 \cdots \int_0^1 f^{\tau_1,\dots,\tau_s}(y_1, \dots, y_s) \prod_{\nu=1}^{s} \left(\{y_\nu - x_\nu\} - \frac{1}{2} \right)^{\tau_\nu} dy_1 \dots dy_s.$$

Note. If r is a positive integer, $\alpha > r + 1$, and $f \in E_s^\alpha(C)$, then we have the following equality analogous to the equality (387):

$$f(x_1, \dots, x_s)$$

$$= \sum_{\tau_1,\dots,\tau_s=0}^{1} \int_0^1 \cdots \int_0^1 f^{r\tau_1,\dots,r\tau_s}(y_1, \dots, y_s) \prod_{\nu=1}^{s} \left[\frac{(-1)^{r-1}}{r!} B_r(\{y_\nu - x_\nu\}) \right]^{\tau_\nu} dy_1 \dots dy_s,$$

where $B_r(x)$ are the Bernoulli polynomials:

$$B_1(x) = x - \frac{1}{2}, \qquad B_2(x) = x^2 - x + \frac{1}{6}, \qquad \ldots .$$

Under $r = 1$ this assertion coincides with (387), and in the general case it is proved by induction with respect to r with the use of the equalities

$$B_r(1) = B_r(0), \qquad B_r'(x) = rB_{r-1}(x) \qquad (r \geqslant 2).$$

THEOREM 39. *Let $r \geqslant 2$ be a positive integer, $\alpha \geqslant 2r$, and $a_1, \ldots, a_s$ be optimal coefficients modulo p. If a function $f(x_1, \ldots, x_s)$ belongs to the class $E_s^\alpha(C)$, then we have the equality*

$$f(x_1, \ldots, x_s) = \frac{1}{p}\sum_{k=1}^{p} \sum_{\tau_1, \ldots, \tau_s = 0}^{1} f^{r\tau_1, \ldots, r\tau_s}\left(\left\{\frac{a_1 k}{p}\right\}, \ldots, \left\{\frac{a_s k}{p}\right\}\right) \prod_{\nu=1}^{s} \left[\frac{(-1)^{r-1}}{r!} B_r\left(\left\{\frac{a_\nu k}{p} - x_\nu\right\}\right)\right]^{\tau_\nu} + O\left(\frac{\log^\gamma p}{p^r}\right), \tag{388}$$

where a constant γ depends on r and s only.

Proof. Let functions $f_1(x_1, \ldots, x_s)$ and $f_2(x_1, \ldots, x_s)$ belong to the classes $E_s^\alpha(C_1)$ and $E_s^\alpha(C_2)$, respectively. We shall show that the product of these functions

$$f_3(x_1, \ldots, x_s) = f_1(x_1, \ldots, x_s) f_2(x_1, \ldots, x_s)$$

belongs to the class $E_s^\alpha(C_3)$, where C_3 depends on C_1, C_2, α, and s.

Indeed, denote by $C_j(m_1, \ldots, m_s)$ $(j = 1, 2, 3)$ the Fourier coefficients of the functions f_1, f_2, and f_3. Multiplying the Fourier series of the functions f_1 and f_2, we obtain

$$f_3(x_1, \ldots, x_s) = \sum_{m_1, \ldots, m_s = -\infty}^{\infty} C_3(m_1, \ldots, m_s)\, e^{2\pi i (m_1 x_1 + \ldots + m_s x_s)},$$

where

$$C_3(m_1, \ldots, m_s) = \sum_{n_1, \ldots, n_s = -\infty}^{\infty} C_1(n_1, \ldots, n_s)\, C_2(m_1 - n_1, \ldots, m_s - n_s).$$

Therefore,

$$\begin{aligned} \left|C_3(m_1, \ldots, m_s)\right| &\leqslant \sum_{n_1, \ldots, n_s = -\infty}^{\infty} \left|C_1(n_1, \ldots, n_s)\, C_2(m_1 - n_1, \ldots, m_s - n_s)\right| \\ &\leqslant \sum_{n_1, \ldots, n_s = -\infty}^{\infty} \frac{C_1 C_2}{\left[\bar{n}_1 \ldots \bar{n}_s (\overline{m_1 - n_1}) \ldots (\overline{m_s - n_s})\right]^\alpha} \\ &= C_1 C_2 \sigma(m_1) \ldots \sigma(m_s), \end{aligned} \tag{389}$$

where $\sigma(m)$ denotes the sum

$$\sigma(m) = \sum_{n=-\infty}^{\infty} \frac{1}{\left[\bar{n}(\overline{m-n})\right]^{\alpha}}.$$

Estimate the sum $\sigma(m)$. If $\overline{m} > 1$, then

$$\begin{aligned}\sigma(m) &= \sum_{|n|\leqslant\frac{1}{2}|m|} \frac{1}{\left[\bar{n}(\overline{m-n})\right]^{\alpha}} + \sum_{|n|>\frac{1}{2}|m|} \frac{1}{\left[\bar{n}(\overline{m-n})\right]^{\alpha}} \\ &\leqslant \frac{2^{\alpha}}{\overline{m}^{\alpha}} \sum_{|n|\leqslant\frac{1}{2}|m|} \frac{1}{\bar{n}^{\alpha}} + \frac{2^{\alpha}}{\overline{m}^{\alpha}} \sum_{|n|>\frac{1}{2}|m|} \frac{1}{(\overline{m-n})^{\alpha}} \\ &\leqslant \frac{2^{\alpha+1}}{\overline{m}^{\alpha}} \sum_{n=-\infty}^{\infty} \frac{1}{\bar{n}^{\alpha}}.\end{aligned}$$

This estimate is, evidently, satisfied under $\overline{m} = 1$ too. But then we get from (389)

$$|C_3(m_1, \ldots, m_s)| \leqslant \frac{C_1C_2}{(\overline{m}_1 \ldots \overline{m}_s)^{\alpha}} \left(2^{\alpha+1} \sum_{n=-\infty}^{\infty} \frac{1}{\bar{n}^{\alpha}}\right)^{s} = \frac{C_1C_2C_4^s}{(\overline{m}_1 \ldots \overline{m}_s)^{\alpha}},$$

and, therefore, $f_3(x_1, \ldots, x_s) \in E_s^{\alpha}(C_3)$, where $C_3 = C_4^s C_1 C_2$ and

$$C_4 \leqslant 2^{\alpha+1}\left(3 + \frac{2}{\alpha - 1}\right).$$

According to the note of Lemma 33 under

$$F(y_1, \ldots, y_s) = f^{r\tau_1, \ldots, r\tau_s}(y_1, \ldots, y_s) \prod_{\nu=1}^{s} \left(\frac{(-1)^{r-1}}{r!} B_r(\{y_\nu - x_\nu\})\right)^{\tau_\nu} \tag{390}$$

the equality

$$f(x_1, \ldots, x_s) = \sum_{\tau_1, \ldots, \tau_s = 0}^{1} \int_0^1 \cdots \int_0^1 F(y_1, \ldots, y_s)\, dy_1 \ldots dy_s \tag{391}$$

holds. Differentiating the Fourier series

$$f(y_1, \ldots, y_s) = \sum_{m_1, \ldots, m_s = -\infty}^{\infty} C(m_1, \ldots, m_s)\, e^{2\pi i\,(m_1 y_1 + \ldots + m_s y_s)},$$

we obtain

$$f^{r\tau_1,\dots,r\tau_s}(y_1,\dots,y_s)$$
$$= C' \sum_{m_1,\dots,m_s=-\infty}^{\infty} m_1^{r\tau_1}\dots m_s^{r\tau_s} C(m_1,\dots,m_s)\, e^{2\pi i(m_1y_1+\dots+m_sy_s)},$$

where $C' = (2\pi i)^{r(\tau_1+\dots+\tau_s)}$. Since

$$\left|m_1^{r\tau_1}\dots m_s^{r\tau_s} C(m_1,\dots,m_s)\right| \leqslant C\frac{|m_1|^{r\tau_1}\dots|m_s|^{r\tau_s}}{(\overline{m}_1\dots\overline{m}_s)^\alpha}$$
$$\leqslant \frac{C}{(\overline{m}_1\dots\overline{m}_s)^{\alpha-r}} \leqslant \frac{C}{(\overline{m}_1\dots\overline{m}_s)^{r}},$$

the function $f_1(y_1,\dots,y_s) = f^{r\tau_1,\dots,r\tau_s}(y_1,\dots,y_s)$ belongs to the class $E_s^r(C_1)$ with the constant $C_1 = |C'|C$.

Let $c(m)$ be the Fourier coefficients of the r-th Bernoulli polynomial $B_r(\{y\})$. Since $c(m) = O(\frac{1}{\overline{m}^r})$, then for the Fourier coefficients of the function

$$f_2(y_1,\dots,y_s) = \prod_{\nu=1}^{s} B_r^{\tau_\nu}(\{y_\nu - x_\nu\})$$

we obtain the estimate

$$C_2(m_1,\dots,m_s) = O\Big(\frac{1}{(\overline{m}_1\dots\overline{m}_s)^r}\Big)$$

and, therefore, the function $f_2(y_1,\dots,y_s)$ belongs to the class $E_s^r(C_2)$. But then the function $F(y_1,\dots,y_s)$ determined by the equality (390) belongs to a certain class $E_s^r(C_3)$ and for the evaluation of the integrals in the equality (391) we may use the quadrature formula obtained under $P = p$ in Theorem 38:

$$\int_0^1\dots\int_0^1 F(y_1,\dots,y_s)\,dy_1\dots dy_s$$
$$= \frac{1}{P}\sum_{k=1}^{P} F\Big(\Big\{\frac{a_1k}{p}\Big\},\dots,\Big\{\frac{a_sk}{p}\Big\}\Big) + O\Big(\frac{\log^\gamma P}{P^r}\Big),$$

where γ depends on r and s only. Hence by (390) we have the equality

$$f(x_1,\dots,x_s) = \frac{1}{P}\sum_{k=1}^{P}\ \sum_{\tau_1,\dots,\tau_s=0}^{1} F\Big(\Big\{\frac{a_1k}{p}\Big\},\dots,\Big\{\frac{a_sk}{p}\Big\}\Big) + O\Big(\frac{\log^\gamma P}{P^r}\Big),$$

which coincides with the theorem assertion by the definition of the function $F(y_1, \ldots, y_s)$.

The interpolation formula (388) is obtained under the assumption that the function $f(x_1, \ldots, x_s)$ belongs to the class $E_s^{\alpha}(C)$, where $\alpha \geqslant 2r$ and $r \geqslant 2$. In the same way, somewhat complicating the proof, we can convince ourselves of the validity of the formula under $r = 1$ also. So if $f(x_1, \ldots, x_s) \in E_s^2(C)$ and $a_1, \ldots, a_s$ are optimal coefficients modulo p, then under $P = p$ we have the equality

$$\begin{aligned} &f(x_1, \ldots, x_s) \\ &= \frac{1}{P} \sum_{k=1}^{P} \sum_{\tau_1, \ldots, \tau_s = 0}^{1} f^{\tau_1, \ldots, \tau_s}\left(\left\{\frac{a_1 k}{p}\right\}, \ldots, \left\{\frac{a_s k}{p}\right\}\right) \prod_{\nu=1}^{s} \left(\left\{\frac{a_\nu k}{p} - x_\nu\right\} - \frac{1}{2}\right)^{\tau_\nu} \\ &\quad + O\left(\frac{\log^{\gamma} P}{P}\right), \end{aligned} \tag{392}$$

where γ depends on s only. Unlike the formula (388), which is not unimprovable, the order of the error decrease in the interpolation formula (392) cannot be improved under any choice of nets.

The quadrature and interpolation formulas with parallelepipedal nets established in this section were obtained under the assumption of the equality $P = p$, where P is the number of the net nodes and p is the modulus of the optimal coefficients. If the quantities $a_1, \ldots, a_s$ are chosen so that the numbers $1, a_1, \ldots, a_s$ are $(s+1)$-dimensional optimal coefficients modulo p, then these formulas are valid under $P < p$ too, but then their precision will be lowered. So, for example, in the formulas (369) and (388) the order of the error decrease will be not $O\left(\frac{\log^{\gamma} P}{P^{\alpha}}\right)$ or $O\left(\frac{\log^{\gamma} P}{P^{r}}\right)$ but $O\left(\frac{\log^{\gamma} P}{P}\right)$ only.

The first results in the application of the number-theoretical nets to the approximate computation of integrals of an arbitrary multiplicity were obtained in the papers [23] and [29]. Henceforward an essential contribution to the number-theoretical methods of numerical integration was made in the articles [3], [12], [14], [10], [8], and [5]. Recently a large number of papers and a series of monographs [30], [15], [18], and [38] have dealt with number-theoretical methods in numerical analysis.

REFERENCES

[1] G. I. Arhipov, *Estimates for double trigonometrical sums of H. Weyl*, Trudy Mat. Inst. Steklov, 142 (1976), pp. 46–66. (In Russian.)

[2] K. I. Babenko, *Fundamentals of Numerical Analysis*, Nauka, Moscow, 1986. (In Russian.)

[3] N. S. Bahvalov, *On approximate computation of multiple integrals*, Vestnik Moskov. Univ., Ser. 1, Mat. Mech., 4 (1959), pp. 3–18. (In Russian.)

[4] D. Burgess, *The distribution of quadratic residues and nonresidues*, Mathematika (London), 4 (1957), pp. 106–112.

[5] V. A. Bykovskii, *On precise order of the error of optimal cubature formulas in spaces with dominant derivative and quadratic discrepancies of nets*, Preprint, Computing Centre of Far-Eastern Scientific Centre of the USSR Academy of Sciences, Vladivostok, 1985, No. 23, 31 p. (In Russian.)

[6] K. Chandrasekharan, *Arithmetical Functions*, Springer-Verlag, Berlin, 1970.

[7] J. van der Corput, *Diophantische Ungleichungen*, Acta Math., 56 (1931), pp. 373–456.

[8] N. M. Dobrovolskii, *Estimates of discrepancies of generalized parallelepipedal nets*, Tula Pedagogical Institute, Tula, 1984, dep. in VINITI 17 01 1985, No 6089. (In Russian.)

[9] T. Estermann, *On the sign of the Gaussian sums*, J. London Math. Soc., 20 (1945), pp. 66–67.

[10] K. K. Frolov, *Upper estimates of the error of quadrature formulas on classes of functions*, Dokl. Akad. Nauk SSSR, 231 (1976), pp. 818–821. (In Russian.)

[11] A. O. Gel'fond, *Differenzenrechnung*, Verlag der Wissenschaften, Berlin, 1958.

[12] J. Halton, *On the efficiency of certain quasirandom sequences of points in evaluating multidimensional integrals*, Number Math., 27, 2 (1960), pp. 84–90.

[13] H. Hasse, *Abstrakte Begründung der komplexen Multiplikation und Riemannsche Vermutung in Funktionkörpern*, Abh. Math. Sem. Univ. Hamburg, 10 (1934), pp. 325–348.

[14] E. Hlawka, *Zur angenäherten Berechnung mehrfacher Integrale*, Monatsh. Math., 66 (1962), pp. 140–151.

[15] E. Hlawka, F. Firneis, and P. Zinterhof, *Zahlentheoretische Methoden in der numerischen Mathematik*, Wien–München–Oldenbourg, 1981.

[16] HUA LOO-KENG, *The additive theory of prime numbers*, Trudy Mat. Inst. Steklov, 22 (1947), pp. 3–179. (In Russian.)

[17] ———, *Die Abschätzung von exponential Summen und ihre Anwendung in der Zahlentheorie*, Teubner, Leipzig, 1959.

[18] HUA LOO-KENG AND WANG YUAN, *Applications of Number Theory to Numerical Analysis*, Springer-Verlag, Berlin–Heidelberg–New York, 1981.

[19] L. A. KNIZHNERMAN AND V. Z. SOKOLINSKII, *On the unimprovability of the A. Weil estimates for rational trigonometric sums and sums of the Legendre symbols*, Moskov. Gos. Ped. Inst., Moscow, 1979, dep. in VINITI 13 06 1979, No 2152. (In Russian.)

[20] N. M. KOROBOV, *Some questions on uniform distribution*, Izv. Akad. Nauk SSSR, Ser. Mat., 14 (1950), pp. 215–238. (In Russian.)

[21] ———, *Distribution of non-residues and primitive roots in recurrent series*, Dokl. Akad. Nauk SSSR, 88 (1953), pp. 603–606. (In Russian.)

[22] ———, *On completely uniform distribution and conjunctly normal numbers*, Izv. Akad. Nauk SSSR, Ser. Mat., 20 (1956), pp. 649–660. (In Russian.)

[23] ———, *Approximate calculation of multiple integrals by number-theoretical methods*, Dokl. Akad. Nauk SSSR, 115 (1957), pp. 1062–1065. (In Russian.)

[24] ———, *On estimation of rational trigonometrical sums*, Dokl. Akad. Nauk SSSR, 118 (1958), pp. 231–232. (In Russian.)

[25] ———, *On zeros of the $\zeta(s)$ function*, Dokl. Akad. Nauk SSSR, 118 (1958), pp. 431–432. (In Russian.)

[26] ———, *On the bound of zeros of the Riemann zeta-function*, Uspehi Mat. Nauk, 13, 2 (1958), pp. 243–245. (In Russian.)

[27] ———, *Estimates of trigonometrical sums and their applications*, Uspehi Mat. Nauk, 13, 4 (1958), pp. 185–192. (In Russian.)

[28] ———, *Estimates of the Weyl sums and distribution of prime numbers*, Dokl. Akad. Nauk SSSR, 123 (1958), pp. 28–31. (In Russian.)

[29] ———, *On the approximate computation of multiple integrals*, Dokl. Akad. Nauk SSSR, 124 (1959), pp. 1207–1210. (In Russian.)

[30] ———, *Number-Theoretical Methods in Approximate Analysis*, Fizmatgiz, Moscow, 1963. (In Russian.)

[31] ———, *Estimates of the sum of the Legendre symbols*, Dokl. Akad. Nauk SSSR, 196 (1971), pp. 764–767. (In Russian.)

[32] ———, *On the distribution of digits in periodic fractions*, Mat. Sb., 89 (131) (1972), pp. 654–670. (In Russian.)

[33] A. I. KOSTRIKIN, *Introduction to Algebra*, Nauka, Moscow, 1977. (In Russian.)

[34] YU. V. LINNIK, *On Weyl's sums*, Dokl. Akad. Nauk SSSR, 34 (1942), pp. 201–203. (In Russian.)

[35] YU. I. MANIN, *On congruences of the third degree to a prime modulus*, Izv. Akad. Nauk SSSR, Ser. Mat., 20 (1956), pp. 673–678. (In Russian.)

[36] L. MORDELL, *On a sum analogous to a Gauss's sum*, Quart. J. Math., 3 (1932), pp. 161–167.

[37] YU. V. NESTERENKO, *On I. M. Vinogradov's mean value theorem*, Trudy Moskov. Mat. Obšč., 48 (1985), pp. 97–105. (In Russian.)

[38] H. NIEDERREITER, *Quasi Monte-Carlo methods and pseudorandoms numbers*, Amer. Math. Soc., 84, 6 (1978), pp. 957–1041.

[39] A. G. POSTNIKOV, *Arithmetic modelling of random processes*, Trudy Mat. Inst. Steklov, 57 (1960), pp. 3–84. (In Russian.)

[40] K. PRACHAR, *Primzahlverteilung*, Springer-Verlag, Berlin–Göttingen–Heidelberg, 1957.

[41] N. N. ROGOVSKAYA AND V. Z. SOKOLINSKII, *On estimation of multiple trigonometrical sums*, Uspehi Mat. Nauk, 40 (1985), pp. 261–262. (In Russian.)

[42] S. A. STEPANOV, *On the number of points of hyperelliptic curve over prime finite field*, Izv. Akad. Nauk SSSR, Ser. Mat., 33 (1969), pp. 1171–1181. (In Russian.)

[43] R. C. VAUGHAN, *The Hardy–Littlewood Method*, Cambridge University Press, Cambrige–New York, 1981.

[44] I. M. VINOGRADOV, *Selected Works*, Springer-Verlag, Berlin–New York, 1985.

[45] ———, *Fundamentals of the Theory of Numbers*, Nauka, Moscow, 1972. (In Russian.)

[46] ———, *A new estimate for the function* $\zeta(1+it)$, Izv. Akad. Nauk SSSR, Ser. Mat., 22, 2 (1958), pp. 161–164. (In Russian.)

[47] A. WALFISZ, *Weylsche Exponentialsummen in der neueren Zahlentheorie*, Verlag der Wissenschaften, Berlin, 1963.

[48] A. WEIL, *On some exponential sums*, Proc. Nat. Acad. Sci. USA, 34, 5 (1948), pp. 204–207.

[49] H. WEYL, *Über die Gleichverteilung von Zahlen mod. Eins*, Math. Ann., 77 (1916), pp. 313–352.

SUBJECT INDEX

INDEX OF NAMES